QGIS FOR HYDROLOGICAL APPLICATIONS

Second Edition

RECIPES FOR CATCHMENT HYDROLOGY
AND WATER MANAGEMENT

HANS VAN DER KWAST
KURT MENKE

LOCATE
PRESS

QGIS 3.22+

Credits & Copyright

QGIS for Hydrological Applications
Second Edition

Recipes for Catchment Hydrology and Water Management

by Hans van der Kwast and Kurt Menke

Published by Locate Press

COPYRIGHT © 2022 LOCATE PRESS
ISBN (PRINT): 978-0986805233
ALL RIGHTS RESERVED.

Direct permission requests to info@locatepress.com or mail:
Locate Press, Suite 433, 113-437 Martin St., Penticton, BC, Canada, V2A 5L1

Editor Gary Sherman
Interior Design Based on Memoir-LATEXdocument class
Publisher Website http://locatepress.com
Book Website http://locatepress.com/book/hyd2

Contents

Foreword

This foreword is addressed directly to you, dear reader, wherever you happen to reside in this geospecial, geospatial world of ours. Whether you are reading these words on a leaf of paper or on a computer screen, flipping to the next page will initiate what is sure to be an enriching journey; and today, I have the distinct privilege of introducing to you the two phenomenal tour guides who will accompany you along the way. The rest of the words in the paragraphs below are perhaps superfluous, because I really have just one statement to offer as you embark on your tour: "You are in good hands with Hans and Kurt!"

You can take my word for it and turn the page now—in which case I wish you well on the road ahead—or, if you're still with me, I'll try to briefly convey why I have so much confidence in your guides.

A few years ago, we at the Australian Water School noticed that many of those who attended our hydrologic and hydraulic modeling courses were unfamiliar with basic GIS terms and would greatly benefit from applying geospatial skills in their modeling. We decided to offer an introductory QGIS course for water modellers and quickly set out to find the most engaging presenters in the industry.

It didn't take long to find our rock stars, Kurt and Hans, who already had a lot of popular content online. Undeterred by time slots that left the Netherlands in the wee hours of the night, they agreed to help us out. As they delivered their courses, we were thoroughly impressed with their professionalism; the accompanying notes and workshop materials were well organised and blended seamlessly with the live, online sessions.

Our first QGIS recording quickly reached 10,000 views. This was before YouTube decided that listing dislikes makes us all grumpy, so we were able to track the impressive fact that there hadn't been a single dislike along the way. Can you imagine 10,000 customers in a row walking past a "How did we do today" screen in a store, and not a single one picks the frowny face? It seems improbable, but based on my own interactions with your hosts, I'm actually not surprised at all. Cool maps, free tech, and straight shooters who know their stuff: What's not to like?

Hans and Kurt are passionate about this community and about making open-source software accessible to the masses, especially in disadvantaged areas. I attended some of their online training sessions that accompany this book during the height of the global COVID lockdowns. Some of the sessions were followed by online "geo-beers", which seemed perfectly designed to help us all keep our sanity while socially distanced. Although these sessions generally stuck Australia with the wee hours, I made a point to open my droopy eyes enough to enjoy the community interactions they facilitated. Given the global spread of QGIS fans, these online forums are continually growing, and this book's authors have played a major role in fostering the underlying community spirit that is driving the growth.

By its very nature, water is geospatial; every drop of it has a place in space on our blue planet. A hydrologist who, by definition, studies water, inevitably serves a dual role as a practicing geospatialologist, if I can coin a very awkward job title. [Perhaps geospecialist or geospatialist would be more palatable terms, but those words haven't made it into the dictionary yet either!] Whichever honorary title you choose to convey upon yourself on completion of the exercises in this book, I'm convinced that you will arrive at the other end with valuable, newfound geo-expertise.

A key element of the hydrological sciences is an understanding of the distribution of water—mapping where it is, where it has been, and where it is going. Communicating that distribution effectively will help you stand out in your field. Whether you are new to the georeferenced world or bringing previous experience to a new platform, this book is sure to provide you with indispensable, time-saving tools that will make your work shine. I have to admit that I still find some occasional zen in manually delineating basin boundaries and flow paths myself; sometimes it's hard to step back and let a machine do all of that work, but the available tools that you will master in this book can sure produce some amazing, high-quality results. [Still, I'd recommend challenging your computer to a little duel on occasion just to stay sharp with your manual skills!]

This book's title rightly calls it a recipe book. But it's a lot more than that; the accompanying plugins and online tutorial videos make this a complete package that is more like getting a recipe book bundled together with a meal-kit delivery service. It's got everything you need to get cooking. If your own cooking skills—like mine—aren't worth beans at the moment, armed with the right ingredients and mentors, you might just be a budding master chef. Bon Appétit!

Krey Price
Director, Surface Water Solutions
Instructor, Australian Water School

About This Book

When I started teaching GIS at IHE Delft Institute for Water Education in 2012, there were not many courses that used open source software. Given that our MSc participants are mostly from the Global South and can't afford expensive licenses, I wanted to change that for my GIS classes. QGIS was the logical alternative, which has all the features my students need for their work in hydrology and water management. In the recent versions, new features have been included that go far beyond what commercial GIS desktop software offers. In 2013, I started teaching QGIS in most of our MSc programmes and in short courses. In 2015, I had a great opportunity to develop new course materials with Jan Hoogendoorn (Vitens) for several trainings for the National Water and Sewerage Corporation (NWSC) in Uganda, funded by Vitens Evides International (VEI) and the IHE Delft Partnership Programme for Water and Development (DUPC). Jonne Kleijer also contributed to the materials during the trainings in Kampala and Lira.

At IHE Delft we had also started our OpenCourseWare platform in 2015. After the trainings in Uganda we agreed with the donors and trainers to make the course materials available as OpenCourseWare with a CC BY-NC license (http://www.gisopencourseware.org). This was an important step enabling many people to learn about QGIS for hydrological applications, even when they were not able to come to IHE Delft for our short courses or MSc modules. The course materials were completed with a YouTube channel (https://www.youtube.com/c/HansvanderKwast) with videos of the lectures and exercises. In the years that followed, I regularly updated the materials following the QGIS Long Term Release (LTR) versions. Many MSc students at IHE Delft inspired me to improve the course materials and add more instructional videos.

In August 2017, I joined a QGIS user conference and hackfest for the first time. This one was organised by Lene Fischer at Skovskolen Forest and Landscape College of the University of Copenhagen in Nødebo (Denmark). It was very inspiring to meet developers of QGIS and to learn about this open source community. Raymond Nijssen introduced me to different ways to contribute to QGIS. It was also here where I met Kurt Menke for the first time. Together with Tim Sutton we worked on the QGIS certification programme and its platform (https://qgis.org/en/site/getinvolved/certification.html). Since Nødebo, I'm part of this great community.

During the short course on QGIS in September 2017, Erik Meerburg (Geo Academie) and I issued the first QGIS certificates. Since then, IHE Delft has issued around 1000 certificates, which is a great way for us to contribute to the further development of QGIS.

In January 2018, I was happy to host the first Dutch QGIS User Group Meeting at IHE Delft, organised in cooperation with Geo Academie. The tracks in Dutch and English attracted participants from diverse backgrounds.

For the short course on QGIS in 2018, I invited Kurt Menke as a guest lecturer to IHE Delft. It was inspiring for our participants to learn about the many new features of QGIS 3.x and how to become part of the community. It was then that the idea for this book was born.

During the QGIS user conference and hackfest in A Coruña (Spain) in March 2019, I followed Kurt's workshop and got inspired to develop a plugin for styling of flow direction maps using the functionality of the Crayfish plugin for styling mesh layers. Together with Radek Pasiok and Peter Petrik from Lutra Consulting, we developed a processing plugin for this during the hackfest. It's available in the Crayfish plugin and works with QGIS 3.6 and later.

For the short course on QGIS for Hydrological Applications in September 2019, I had invited Kurt again. We had launched the first edition of this book, which has been used in many courses since then.

Since the first edition of this book, many updates in QGIS have shortened and improved the work flows in this book and added great new styling features. In addition, I was happy to launch the PCRaster Tools plugin, developed together with Nyall Dawson, in September 2021. This processing provider plugin adds approximately 100 operators for map algebra to the Processing Toolbox. These improved work flows and the PCRaster Tools plugin are included in the second edition of this book.

With this book and the courses, I hope that many more GIS professionals get interested in using QGIS for hydrological applications and become part of the QGIS community.

Last, but not least, it is important to mention that by purchasing this book you are supporting students to attend FOSS4G and QGIS events. In this way, I hope to contribute to increasing the diversity in the open source GIS community.

IHE Delft Institute for Water Education

IHE Delft Institute for Water Education (http:/www.un-ihe.org) is the largest international graduate water education facility in the world and is based in Delft, the Netherlands. The Institute confers fully accredited MSc degrees, and PhD degrees in collaboration with partner universities. Since 1957, the Institute has provided water education and training to 23,000 professionals from over 190 countries, the vast majority from Africa, Asia, and Latin America. Also, numerous research and institutional strengthening projects are carried out in partnership to strengthen capacity in the water sector worldwide. Through our overarching work on capacity development, IHE Delft aims to make a tangible contribution to achieving all Sustainable Development Goals in which water is key.

Since 2013, QGIS is the default in most MSc programmes. From October 2019, all MSc students learn the concepts of GIS with QGIS in the first modules. Besides GIS education in the MSc modules, IHE Delft offers short courses, online courses, OpenCourseWare and tailor-made trainings, where the use of open source software (QGIS, Python) and open data is key.

Who This Book is For

The book is designed for professionals active in hydrology and water management, especially those involved in using simulation models for water management and performing GIS analysis. However, the book is also suitable for beginners in QGIS who want to learn GIS concepts and QGIS features in a problem based learning manner. The book can be used with an instructor in the classroom, or as independent study for beginners and experts. Each chapter introduces a case and concludes with styling recipes introducing the many features that QGIS offers. These concluding styling sections provide a solid foundation in the robust cartographic capabilities found within QGIS. Topics such as inverted polygon shapeburst fills, advanced label settings, and blending modes are covered.

The first chapter covers georeferencing and digitizing of vectors. In the second chapter, the reader learns how to import spreadsheets, join and edit attribute tables, and interpolate point data to raster. Chapter 3 covers map algebra for spatial analysis. Different raster data types (boolean, discrete, and continuous) are used in calculations, such as proximity analysis, reclassification, and Boolean logic. The Raster Calculator and functions in the Processing Toolbox are also introduced.

The most common use of GIS for professionals in hydrology and water management is covered in Chapter 4, where the reader is guided through the procedure to perform stream and catchment delineation. Chapter 5 proceeds with showing how to add open data from web map services to the delineated catchment. Chapter 6 was inspired by my MSc thesis students who needed percentages of land cover for each subcatchment. While going through the steps, this chapter will also introduce vector geoprocessing functions such as intersect and dissolve. The chapter concludes with using the Data Plotly plugin for making pie charts. The book concludes with designing a well-crafted map in the print layout.

The Data

The data for this book are available for download at `http://locatepress.com/qgis_hydrological` (202 MB). All the data are in the public domain and most of it can be downloaded from open data portals that are introduced in this book. The book also contains links to videos on theory and practice with QGIS.

About the Authors

Hans van der Kwast is senior lecturer at IHE Delft. He received a Master's degree in Physical Geography at Utrecht University in the Netherlands in 2002, with a specialization in GIS and Remote Sensing. In 2002, he was appointed at the Faculty of Geosciences of Utrecht University as a junior lecturer in GIS. In 2009, he defended his PhD at Utrecht University on the integration of remote sensing in spatial dynamic modeling of soil moisture using data-assimilation techniques implemented in the PCRaster Python framework. From 2007 to 2012, he worked at the Flemish Institute for Technological Research (VITO) where he was appointed as a researcher in spatial dynamic environmental modeling. He participated in projects related to water quality modeling, land-use change modeling, and the use of remote sensing for urban applications. As a guest lecturer, he supervised BSc students in a hydrological and geomorphological fieldwork as part of the curriculum Earth Sciences at Utrecht University. Since April 2012, he works at IHE Delft. In his teaching, he actively promotes the use of Free and Open Source Software (FOSS) by young professionals from the Global South. He is also involved in many research and capacity development projects related to GIS, remote sensing, spatial data infrastructures, Python, open data, and citizen science. He is a QGIS Certified Instructor and board member of the Dutch QGIS User Group.

A former archaeologist, Kurt Menke is a geospatial generalist based out of Helsinge, Denmark. Kurt Menke has a broad skill set. He is a spatial analyst, cartographer, web map developer, trainer/teacher, and author. He received a Master's degree in Geography from the University of New Mexico in 2000. His main areas of focus are public health, conservation, and education. In 2021, he accepted a position with Septima P/S and moved to Denmark with his wife Sarah.

Kurt has a long history using QGIS, first downloading version 0.7 (Seamus) in 2005. He is an open source GIS authority, having co-authored three editions of Mastering QGIS for Packt Publishing and authoring *Discover QGIS 3.x* (`https://locatepress.com/dq3`) through Locate Press (May 2019). He can frequently be found speaking at FOSS4G and QGIS conferences. In 2015, he became an OSGeo Charter Member. He is an experienced FOSS4G educator and is a co-author of the GeoAcademy (U.S. version: `https://github.com/FOSS4GAcademy`). He is now a QGIS Certified Instructor. His offerings range from a semester long Intro to FOSS4G course he originally developed in 2009, to short courses and professional workshops. In partnership with the U.S. National Library of Medicine (NLM) he created a program named Community Health Maps (`https://www.communityhealthmaps.org`). This program aims to empower underserved and minority populations with open source mapping technology. Kurt has trained over 800 public health workers in this workflow over the last five years. In 2015, he was awarded the Global Educator of the Year Team Award by GeoForAll as part of the GeoAcademy team.

Acknowledgments

Hans would like to thank IHE Delft and his colleagues for supporting him in developing high-quality curricula with open source GIS for their MSc students. He's very grateful to have the chance every year to introduce around 200 MSc, PhD, and short course participants from the Global South to open source software. He would also like to thank these participants who inspire him to keep up the good work and teach the state-of-the art. Hopefully, we can make a change and get rid of the vendor lock-in found in many countries.

In addition, he would like to thank the donors and project partners who supported the development of course materials with QGIS, enabling their beneficiaries to learn open source alternatives.

He would like to thank Kurt Menke for being his co-author. It's great to work with Kurt. Without him, the layer styling sections and map design chapter would not have been of exceptionally high quality.

Kurt is grateful to Hans van der Kwast for inviting him to partner on this book. After being invited as a guest lecturer at IHE Delft in 2018—teaching this same material—it's exciting to see it come to fruition in book form. He would like to thank the QGIS community for constantly sharing knowledge and inspiring him to use new techniques in the ever evolving QGIS project.

We would both like to thank Krey Price for his thoughtful foreword.

And of course, a huge thanks to the founder of Locate Press, Tyler Mitchell. His ongoing support has been invaluable. He has also infused Locate Press with a ton of energy including social media promotion, a rebranding with a new LP logo. This is very much appreciated by us authors. Finally, we would like to thank Gary Sherman, the founder of QGIS and previous owner of Locate Press. He is always a keystroke away, helping us out with questions about LaTEX, RST, and git.

A note from Hans: I'm writing this section at the same location where I did my PhD research: the rural community of Sehoul in Morocco. A nice place for a writer's retreat but moreover a place where I keep my feet on the ground to keep in my mind why the skills in this book are so important: to improve catchment management and protect vulnerable areas and people from floods, droughts and environmental problems.

1. Preparing Data from Hard Copy Maps

1.1 Introduction

In order to use hard copy maps in a GIS, they need to be scanned and georeferenced. Georeferencing is also needed for raw remote sensing images, such as aerial photographs and satellite images.

For the best result, choose a map sheet that is clean and does not have too many folds. Use a scanner that is large enough to scan the whole map. The resolution of the scanner should be high enough (e.g. 1200 dpi) to have sufficient detail in the resulting raster maps.

For georeferencing we need to link locations on the scanned image to coordinates. There are two methods:

- Collect ground control points (GCPs) at locations that are clearly visible in the image, such as bridges and junctions. GCPs can be collected with a GPS in the field or from another GIS layer.
- If the hard copy map contains a coordinate grid, use the printed grid as a reference. Make sure that you know the projection of the grid, which is usually stated on the map.

After this chapter you will be able to:

- find the projection and *EPSG* code of a map
- install plugins
- *georeference* a scanned map using *GCPs* from a grid
- use the *coordinate capture tool*
- use online layers from the *QuickMapServices* plugin
- digitize points, lines, and polygons and add attributes
- use the *snapping toolbar*
- *dissolve* features
- store data in a *GeoPackage*
- style and label vectors

Figure 1.1, on the following page shows the workflow of this chapter.

In this chapter we will use a scanned map of Mount Marcy (USGS, 1979), Mount_Marcy_New_York_USGS_topo_map_1979.JPG, which we will georeference with the coordinate grid printed on the map. You can download the file from http://locatepress.com/qgis_hydrological. Save it to a local folder.

> For this chapter, the videos from the *QGISHydro Chapter 1* playlist at YouTube channel http://www.youtube.com/c/hansvanderkwast provide the theoretical background, results of the steps, and additional materials.

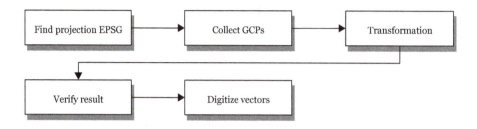

Figure 1.1: Workflow for Georeferencing a Scanned Map and Digitizing Vectors

1.2   Choosing the Projection

Using any image viewing software, have a look at the scanned map and try to determine the projection. At this stage don't open the picture in QGIS yet.

- Which projection was used?
- Look for the EPSG code at http://www.spatialreference.org and write it down.

1.3   Importing the Scanned Map into the Georeferencer

1. Start QGIS Desktop

2. From the main menu choose Raster | Georeferencer (figure 1.2).

Figure 1.2: Starting the Georeferencer from the Raster Menu

The Georeferencer is now opened in a new window.

3. Maximize this window.

4. Click the *Open Raster* button

5. Browse to the Mount_Marcy_New_York_USGS_topo_map_1979.JPG file. A pop-up window might appear where you have to specify the *Coordinate Reference System (CRS)* of this input map. It does not yet have a CRS, so you can click *Cancel*.

1.4   Setting the Transformation Parameters

First we have to set the transformation settings.

1. In the georeferencer menu choose Settings | Transformation Settings; here you can choose:

- Different *transformation types*. A simple linear transformation can be used if the map is not much deformed. The other ones can be tried when more deformation exists. More information about the transformation types can be found in the QGIS documentation `https://edu.nl/tc3ae`. We will start with a *Linear transformation*.
- *Resampling method*: If you need the pixel values in further calculations, it is best to choose the nearest neighbor option. This resampling method will preserve as much as possible the original pixel values by choosing the nearest one. Visually, however, this method results in a "blocky" map. If the purpose is only for visual use, for example as a backdrop for digitization of vector layers, then it is better to choose another resampling method. Here we will use the *Cubic* method, which uses the average of the 16 nearest pixels.
- *Target SRS*: Choose the code that you noted before: EPSG:26718. You can choose it by clicking ⬕ and typing `26718` in the filter field.

2. Browse to the folder where you want to save the georeferenced map using the ⬕ button. The tool automatically adds `_modified` to the file name. In our case, the georeferenced file will be named `Mount_Marcy_New_York_USGS_topo_map_1979_modified.tif`. Keep the other settings on default and make sure the box *Load in QGIS when done* is checked.

The dialog should look like figure 1.3, on the following page.

3. Click *OK* to continue.

1.5 Adding Ground Control Points (GCPs)

In order to link the file coordinates to real world coordinates, we need to indicate Ground Control Points (GCPs) with known coordinates. We can derive these coordinates in different ways:

- The easiest way is to use the coordinate grid on the scanned map if this is available and if it is in a known projection. We click on a node in the grid and type the corresponding X and Y coordinates in the dialog.
- Using a reference map in the QGIS map canvas that has already been georeferenced. In this way we can obtain the right coordinates by clicking on the reference map.
- Using GCPs that were measured in the field using a GPS.

Here we will use the coordinate grid that is printed on the map.

1. Zoom in on the node with coordinate 581000 East and 4885000 North. Read the coordinates from the side of the map to determine the X and Y values.

2. Click the *Add Point* ⬕ button to add a GCP.

3. Enter the X and Y coordinates in the pop-up window (figure 1.4, on the next page).

> If you have a reference map in the QGIS map canvas, you can use the *From Map Canvas* button to capture the coordinate and perform an image to image georeferencing.

4. Press *OK*.

Now your screen should look like figure 1.5, on page 17.

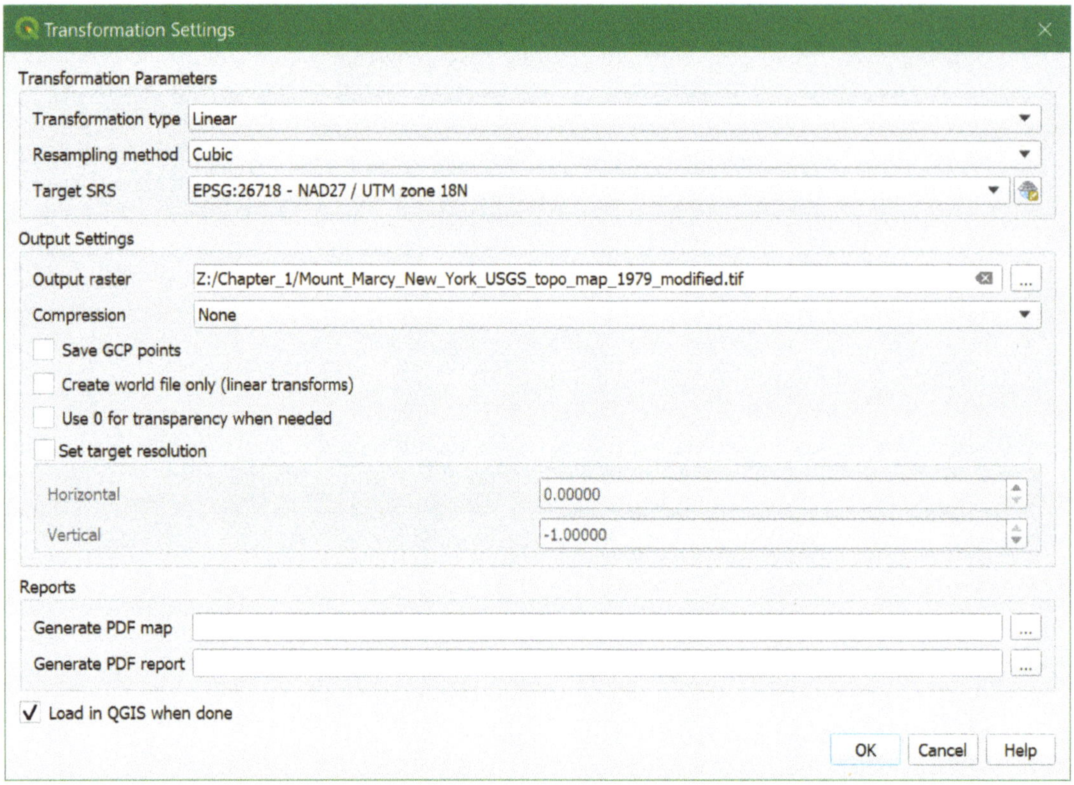

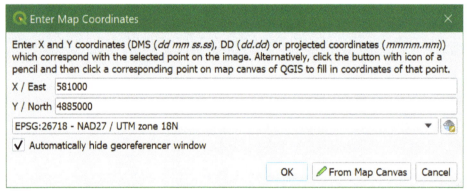

Figure 1.3: Transformation Settings Dialog

Figure 1.4: Enter Map Coordinates

The red dot is the location that you have referenced. In the table below the map, you can see the *Source X* and *Source Y* coordinates. These are the unreferenced file coordinates. Their values depend on which pixel you clicked for placing the GCP, so it can differ from the screenshot in figure 1.5. *Dest. X* and *Dest. Y* show the real world coordinates that you have linked to this location. The other fields of the table have to do with estimated accuracy and will be filled in after adding more points.

5. Let's choose a second GCP in the upper right corner of the map using the same method as we used for the first GCP. Your screen should look like figure 1.6, on the next page.

You can see that some error statistics have been calculated. With only two points this does not make much sense. The minimum amount of GCPs for a linear transformation should be

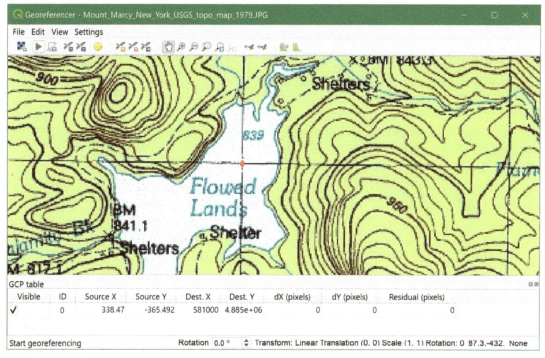

Figure 1.5: First GCP

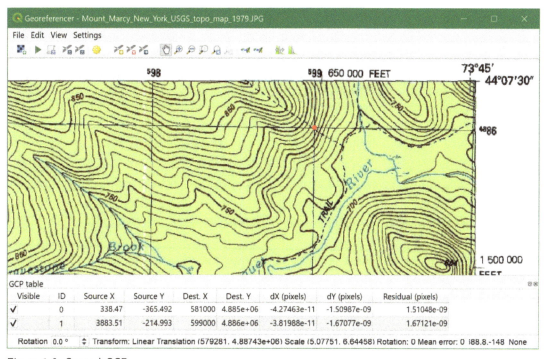

Figure 1.6: Second GCP

three. With more GCPs that are well distributed over the map, the transformation will be more accurate.

6. Using the same method, add a GCP in the lower left and lower right corners of the map. If you make a mistake, you can remove the GCP using the *Delete point* button ![delete point icon] . Your screen should look like figure 1.7.

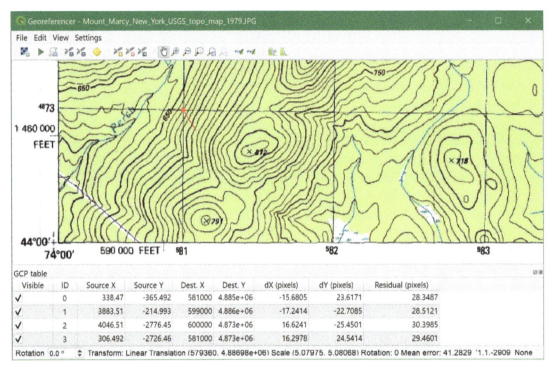

Figure 1.7: Four GCPs added

1.6 Reduce Errors and Perform the Transformation

At the bottom of the screen you can see the estimated mean error (41.2829 pixels in our case). The error is also visualized at the GCPs using a red line. Obviously your values will not be exactly the same.

There are two ways to reduce the error:

- You can use the *Move GCP Point* button ![move gcp point icon] to place the GCPs really at the nodes where the grid lines cross. You need to zoom in to select the right pixel. Note that the mean error is not automatically updated. You need to change the transformation type to something else and then back.
- If the first option doesn't reduce the error, we can change the transformation type. If we change to another transformation type in the transformation settings, the error values will be recalculated.

In this exercise we'll apply the second option after we have checked that are GCPs have been placed correctly.

1. In the menu, go again to Settings | Transformation Settings or click the gear icon.

2. Now let's select a first order polynomial (*Polynomial 1*) instead of the linear transformation. Keep the rest as is.

3. Click *OK* to return to the GCP table.

Now you can see that the mean error has been reduced to a fraction of a pixel, which is acceptable. If you don't see a mean error < 1 pixel, you will have to check the GCP locations and correct them.

> A mean error of < 1 pixel cannot always be achieved. The decision to accept a certain accuracy depends on the purpose of the map.

4. Now we can start georeferencing using the ▶ button.

After some calculation time you will see a pop-up (figure 1.8).

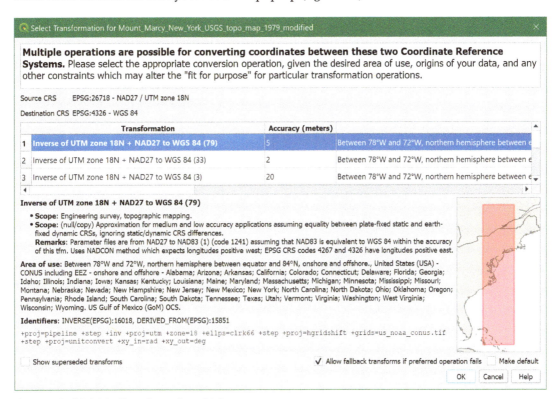

Figure 1.8: Multiple Transformations Dialog

This means that there are multiple transformations possible. In this book we will always use the default one. Only use the others if you need a different accuracy for specific applications.

5. Click *OK*.

The georeferenced map appears in the QGIS map canvas, which is in another QGIS window.

6. You can close the *Georeferencer* window. It will ask if you want to save your GCPs. You can click *Discard* if you don't want to use them. If you save them, you can load them again in the *Georeferencer*.

1.7   Verify the Georeferenced Map

You can verify the result by checking the coordinates at the grid nodes that you used as GCPs.

1. Right-click on a grid node in the map canvas to display the coordinates next to the cursor (figure 1.9).

Figure 1.9: Copy Coordinates

The *Map CRS* shows the coordinates in the projection of the project (i.e. EPSG:26718), the second field shows the coordinates in the Geographic Coordinate System (i.e. EPSG:4326), with coordinates in decimal degrees of longitude and latitude.

2. Read the coordinates from the side of the map and verify that they correspond to the coordinates at the location where you clicked. Repeat this for all GCP nodes.

Another way to verify the result is to use a web map as a backdrop. The *QuickMapServices* plugin provides easy access to many web maps such as Google Satellite and OpenStreetMap.

Plugins are third-party additions to QGIS. They can be installed through the *Plugins manager*. You need an internet connection to connect to the *Plugins repository*.

3. From the main menu, choose Plugins | Manage and Install Plugins....

4. Search for *QuickMapServices* (figure 1.10, on the next page).

5. Click *Install Plugin*. Click *Close* after installing.

6. In order to get access to more online resources choose Web | QuickMapServices plugin | Settings from the main menu. Next, choose the *More services* tab and click *Get contributed pack* (figure 1.11, on the facing page). Click *OK* in the pop-up and *Save* in the dialog when the contributed pack is installed.

7. Go to the main menu and choose Web | QuickMapServices plugin and select Google | Google Terrain and then OSM | OSM Standard. Compare the georeferenced map with Google Terrain and OpenStreetMap.

To help with this comparison you will employ a *Blending Mode*. Blending modes determine how two layers interact visually. When a blending mode is applied to a layer it will be blended with the layer below.

8. Open the *Layer Styling panel* by clicking the 🖌 button. Set the target layer to

Mount_Marcy_New_York_USGS_topo_map_1979_modified.

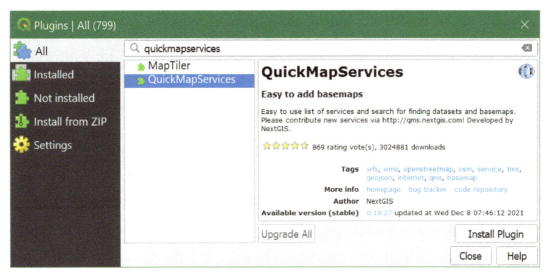

Figure 1.10: Installing the QuickMapServices Plugin

Figure 1.11: QuickMapServices Settings to Install More Services

9. In the panel, choose *Blending mode Multiply* (figure 1.12, on the next page).

> Blending modes allow for more elegant rendering between GIS layers. They can be much more powerful than simply adjusting layer opacity. Blending modes allow for effects where the full intensity of an underlying layer is still visible through the layer above. There are thirteen blending modes available. A nice overview can be found in this blog by Helen McKenzie: https://loc8.cc/hyd/blendmodes

10. After comparing the georeferenced map with the online layers, change the blending of *Mount_Marcy_New_York_USGS_topo_map_1979_modified* back to normal. We are going to use this layer in the next sections as a backdrop to digitize vectors. You can uncheck the *Google Terrain* and *OSM Standard* layers.

This is a good time to save your QGIS project (.qgz file).

11. From the main menu, choose Project | Save as....

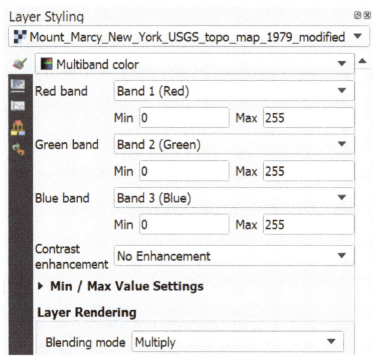

Figure 1.12: Setting the Blending Mode to Multiply

The project file contains references to all layers (not the data itself), styles, projections, extent and zoom level of the map canvas. Save your project regularly!

Try to use the Georeferencer to register the scanned map to a satellite image from the QuickMapServices plugin. In this way you can perform an image to image registration. Please make sure you use the right projection. Does the image to image registration give better results?

1.8   Digitizing Vector Layers from a Georeferenced Backdrop

Our georeferenced scanned map can now be used as a backdrop to digitize vector layers. Vectors can be points, (poly)lines or polygons. In this exercise we are going to digitize:

- Mountain tops as points
- Rivers as polylines
- Lakes as polygons

We will create these vector layers in a GeoPackage spatial database. In that way we have all layers together in one file, instead of using separate shapefiles.

Digitize Peaks

First we have to create an empty GeoPackage layer.

1. From the main menu, select Layer | Create Layer | New GeoPackage Layer.... You can also use the *New GeoPackage Layer* button, or the keyboard shortcut for your platform (e.g.

`Ctrl + Shift + N` on Windows).

2. In the *New GeoPackage Layer* dialog, first select the *Database* filename by clicking the ... button. Browse to the folder where you want to store the GeoPackage and save it to `Mount_Marcy.gpkg`. For *Table name* choose `Peaks`. For *Geometry type* choose *Point*. Make sure `EPSG:26718` is chosen as the projection. Create a new field with the *Name* `Elevation`, *Type* `Whole Number (integer)` and click [🔲 Add to Fields List] (figure 1.13).

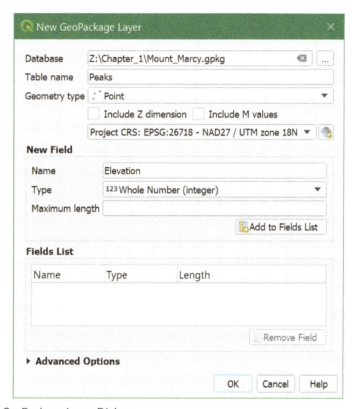

Figure 1.13: New GeoPackage Layer Dialog

3. Add a second new field with *Type*: `Text` named `Name`. Click *OK*.

The empty layer has now been added to your layers list (figure 1.14).

Figure 1.14: Peaks Added to the Layers List in the Layers Panel

In order to start digitizing, you have to toggle editing mode on.

4. Click on the *Peaks* layer so it is selected. Click on 🖉 to toggle editing mode. You can also right-click on the layer and choose *Toggle Editing* from the context menu to place the layer into edit mode.

Now the buttons on the *Digitizing Toolbar* become active. A pencil icon also appears next to the layer in the *Layers* panel indicating the layer is in edit mode.

5. On the topographical map navigate to a mountain. If a geodetic datum is present on top there will be an X and an elevation value.

6. When you have found one, zoom in and click the *Add Point Feature* button . The cursor changes to a crosshair. Move the mouse to the mountain top. Click on the mountain top.

A dialog with a form shows up. *fid* is the feature id that is automatically generated. It's a unique integer number for each feature. The other attributes that we have to fill in are Elevation and Name.

7. In this example we type 1501 and Mt Skylight for *Elevation* and *Name* respectively (figure 1.15). If you map a peak without a labeled elevation determine the elevation based on the contour lines.

Figure 1.15: Add Feature Attribute of Peaks

8. Repeat this step for a few other peaks.

If you made a mistake, you can use the *Vertex Tool* to move the feature or use one of the select options (figure 1.16) to select the point feature and click to delete the selected point feature. These buttons can be used to undo/redo editing actions. Use to save the edits.

Figure 1.16: Select Options

9. When done, click on the button to toggle editing off. If you didn't save edits yet, it will ask you to *Save* or *Discard*. With *Discard* you can always undo your edits until the last time it was saved.

10. You can check the attribute table of your new point vector layer by right-clicking on the layer name (*Peaks*) and selecting Open Attribute Table (figure 1.17).

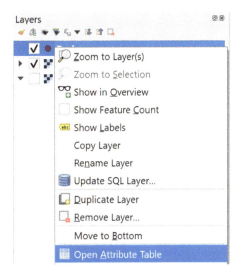

Figure 1.17: Open Attribute Table

Now you can see the attributes that you have added and their *fid*, *Elevation*, and *Name* values (figure 1.18).

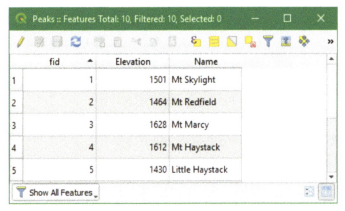

Figure 1.18: Attribute Table of the Peaks Layer

Digitize Rivers

Your next task is to digitize line features (rivers). The procedure is similar to creating a point layer.

11. In the *New GeoPackage Layer* dialog first browse to the existing *Database*, Mount_Marcy.gpkg.

12. Give a new *Table name* of Rivers and a *Geometry type* of LineString. As a new attribute we add Name with the type *Text data*. Don't forget to click the Add to Fields List . Check if the dialog resembles figure 1.19, on the following page and click *OK*.

A *New GeoPackage Layer* window will open informing you that the file already exists.

13. Choose *Add New Layer* (figure 1.20, on the next page).

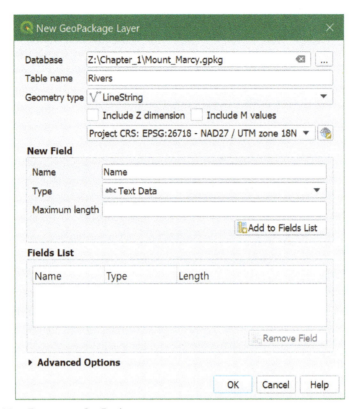

Figure 1.19: Add Line Feature to GeoPackage

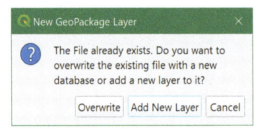

Figure 1.20: Add a New Layer to the GeoPackage

The *Rivers* layer is now added to the layers panel.

14. Toggle editing to start adding rivers. Click on the *Add Line Feature* button to add a new river feature. Zoom and pan on the map to find a stream to digitize.

> Always start digitizing from the upstream to downstream. The direction will be stored in the layer. Always place a vertex when a tributary joins a larger stream. That's important when connecting the tributary later.

15. Click on the starting point of the line (node) and click when necessary to make a vertex.

You can use the zoom and pan buttons to trace the stream. You can use the spacebar to pan during digitizing. With Backspace you can delete the last node while digitizing.

16. After you place the end node of the line, right-click to complete the feature. Now you can enter the attribute data in the form (figure 1.21).

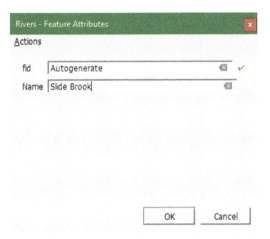

Figure 1.21: Add Line Feature Attributes

Now you are going to digitize a tributary. First you have to set the snapping options.

17. In the main menu choose View | Toolbars | Snapping Toolbar (alternatively right-click on a toolbar and choose Snapping Toolbar). Click <img> to enable snapping. Choose to snap to vertices of the active layer with a tolerance of 15 meters (figure 1.22).

Figure 1.22: Snapping Toolbar

18. Now digitize the tributary from upstream to where it joins the higher order river. You will see that the line will snap to the node that you placed before on the main river.

If you want all tributaries to be one single feature, you need to *dissolve* the features.

19. If you have given all tributaries the same name, you can dissolve them by choosing Vector | Geoprocessing Tools | Dissolve from the menu. Use the ... button to choose Name as the *Dissolve field* and save the result in a GeoPackage layer with the name Rivers_dissolved. Click *Run* (figure 1.23, on the following page).

20. Check the attribute table of the result (figures 1.24 and 1.25, on the next page).

Digitize Lakes

Finally we are going to create a polygon vector layer for some lakes. Try to find out how to do this for yourself. It is very similar to the procedure for lines. The only difference is that the first node should be the same as the last node in order to close the polygon. Use the name of the lake as a text attribute.

Some lakes have islands that need to be removed from the lake features. For this purpose, enable the *Advanced Digitizing Toolbar* in a similar way as we enabled the *Snapping Toolbar* earlier. Click <img> and digitize the island in the same way as you digitize a polygon. After finalizing the ring, it will be removed from the lake polygon.

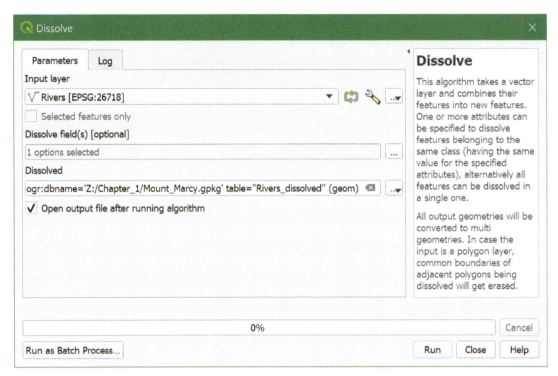

Figure 1.23: Dissolve Dialog

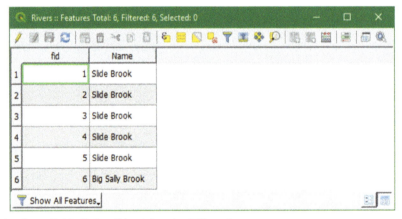

Figure 1.24: Attribute Table Before Dissolving the Tributaries

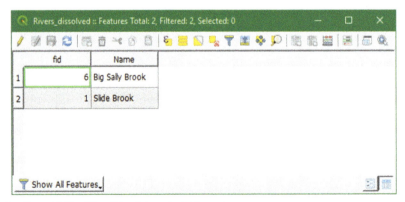

Figure 1.25: Attribute Table After Dissolving the Tributaries

The order of the layers is also important. The order of visualization in the map canvas follows the order in the Layers panel. The first layer is on top. The general order of layers is: points, lines, polygons, rasters. Vectors below rasters are not visible, unless you use blending or opacity. This however changes the colors and should only be used if it has a purpose. You can deviate from the general order if that makes the visualization clearer. You can change the order of layers by dragging the layers in the Layers panel to the desired position.

1.9   Styling the Mountains, Rivers, and Lakes

In this final section you will style your peaks, rivers, and lakes.

Style Peaks

You will begin with the Peaks layer.

1. Open the *Layer Styling panel* by clicking the 🖌 button. Set the target layer to *Peaks*.

By default they are styled using a *Simple marker*.

2. Select *Simple marker* and change the *Symbol layer type* to *SVG Marker*. Below you will find a section named *SVG Groups*. Here you can browse for SVG icons that are installed with QGIS. Find the *symbol* folder and click on it. Scroll down to find the *red marker* 📍 icon and select it. Change the *Width* and *Height* to 12 mm each. See figure 1.26.

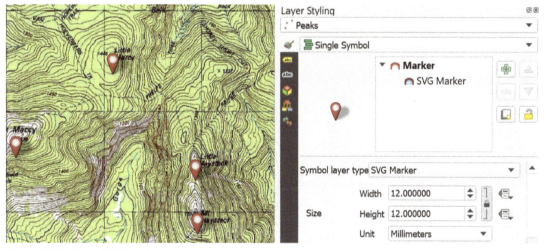

Figure 1.26: Peak SVG Symbols

Label Peaks

Next you will label the peaks.

3. Switch to the *Labels* tab ⟨abc⟩ of the *Layer Styling Panel*. Switch from *No Labels* to *Single labels*. Set the *Label with* option to the Name field. It is also possible to use multiple fields in a feature label by using an expression.

4. Click the *Expression* $\mathcal{E}$ button to open the *Expression Dialog* window. Expand the *Fields and Values* section and add the Elevation field after the Name field. When combining text elements in an expression they need to be separated by the *String Concatenation* || operator.

5. Additionally, the *New line* '\n' operator can be used to wrap the new column onto a second line. However, it requires another *String Concatenation* operator after it. Set up an expression like this: "Name" || '\n' || "Elevation". This is nice, but it may not be clear what the number represents and the labels will be easier to read with a thousands separator.

6. To add units of measure to the elevation, you can add the string *meters* after the value by appending || ' meters' to the existing expression (note the space before meters). To accomplish the second enhancement you will use the format_number function. Use the search box to find the format_number function. Insert it right before the Elevation field. The help panel will show you the syntax for this function. It requires a number and a number of decimal places. The number will be the Elevation field and the number of places 0. This will simply format the data to a number and insert a thousands separator. See figure 1.27.

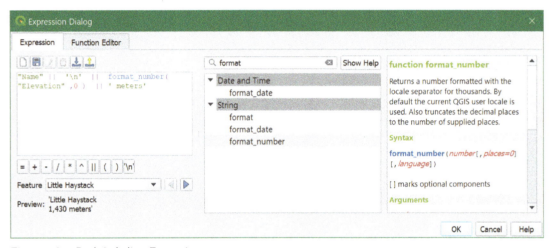

Figure 1.27: Peak Labeling Expression

7. To make the labels easier to read change the font *Style* to Bold. Switch to the *Label buffer* tab **abc** and check the *Draw text buffer* option. To give more separation between the labels and the feature icon switch to the *Label placement* tab and set the *Distance* to 2 mm.

8. Finally click the *Automated placement settings* button to open the *Automated Placement Engine* window. Uncheck the box for *Allow truncated labels on edges of map* option. This will prevent labels from being cut off.

Your labels should look like figure 1.28, on the facing page.

Style Rivers

Next you will style the rivers.

9. Set the target layer in the *Layer Styling Panel* to *Rivers* and switch from the Labeling tab back to the Styling tab.

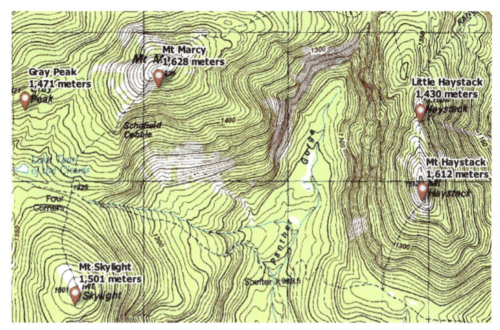

Figure 1.28: Peak Labeling

10. Select *Simple line*. Click the *Color* bar to open the *Select Line Color* panel.

11. Change the *Color* to an RGB value of 31|120|180.

12. Click the *Go back* ◀ button to return to the main symbology panel.

13. Increase the *Stroke width* to 0.86 mm.

14. To label the rivers switch to the *Labels* tab ⟨abc⟩ .

15. Repeat the initial steps of labeling Peaks to label rivers with just the Name field.

16. Switch to the *Label placement* ✥ tab and choose for *Mode* the *Curved* option.

To make them more readable against the topo map you will apply a buffer.

17. Switch to the *Label buffer* tab **abc** and check *Draw text buffer* option.

You will set the color to the green background of the topo map.

18. Click the drop-down arrow for the *Color* setting and choose *Pick color*. With the eye dropper cursor click on a place to select that green topo map background color.

19. Finally return to the *Text* tab **abc** and set the label *Color* to an RGB value of 31|38|180, the *Font* to Calibri, the *Size* to 11 points and the *Style* to *italic*.

> If your computer doesn't have a *Calibri* font, choose a suitable sans serif font to label your features.

Your rivers should look like figure 1.29, on the next page.

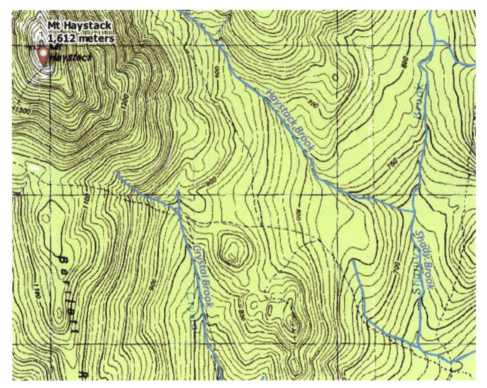

Figure 1.29: River Styling

Style Lakes

Next you will style the lakes. You will use a shapeburst fill which will allow you to color the lakes from light blue to dark blue.

20. Set the target layer in the *Layer Styling Panel* to *Lakes* and switch from the *Labeling* tab back to the *Styling* tab.

21. Select the *Simple fill* styling component. Change the *Symbol layer type* to *Shapeburst fill*. Keep the default *Gradient colors* setting of *Two color*. Set the first color to an RGB value of 185|239|255. Set the second color to an RGB value of 31|133|180.

22. Set the *Shading style* to *Set distance* with a value of 6. Increase the *Blur strength* to 12.

Finally you will add a simple line to represent the coastline of each lake.

23. Click the *Add symbol layer* button. Change the new *Simple fill* renderer to a *Symbol layer type* of *Outline: Simple line* with *Color* of 31|133|180.

24. Label the lakes with the the Name field. Set the label *Color* and RGB value of 185|239|255 (light blue), the *Font* to Calibri, the *Size* to 10 points and the *Style* to *bold italic*.

25. Switch to the *Label placement* tab and change the *Mode* to *Horizontal*. Then switch to the label *Formatting* tab and enter a space as the *Wrap on character*. Set the *Alignment* to *Center*.

Your lakes should look like figure 1.30.

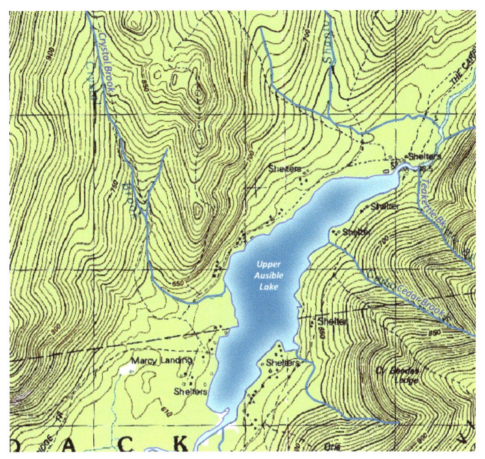

Figure 1.30: Lake Styling

2. Importing Tabular Data into QGIS

2.1 Introduction

Often we get data in tabular format, for example spreadsheets or CSV files. Sometimes the data comes in two tables—one with the coordinates and another one with the attributes you need for your analysis.

After this chapter you will be able to:

- *import tabular data* into a GIS
- save tables with geometry to GIS format
- *join attribute tables*
- *edit attribute tables*
- *interpolate* point data to raster
- style and label point vectors
- style continuous rasters

In this example we will import a table with the daily average temperature on September 1, 2013 at several meteorological stations in the Netherlands. The data was downloaded from the KNMI Data Centre (Royal Netherlands Meteorological Institute, http://data.knmi.nl), but reformatted for the purpose of this exercise.

Figure 2.1 shows the workflow of this chapter.

Figure 2.1: Workflow for Importing Spreadsheets, Joining Attributes and Spatial Interpolation

In this exercise we'll use the following data:

- KNMI_20130901_tday.xls: table with average daily temperatures for different stations
- KNMI_stations.xls: table with station number and coordinates of the location of the stations

This data can be downloaded from http://locatepress.com/qgis_hydrological.

For this chapter videos from the *QGISHydro Chapter 2* playlist at YouTube channel http://www.youtube.com/c/hansvanderkwast provide the theoretical background, results of the steps and additional materials.

2.2   Check the Spreadsheets

It is good practice to check the contents of spreadsheets or delimited text files before importing them into any software. For example, when importing CSV files it is important to know the column separator.

Open the files KNMI_20130901_tday.xls and KNMI_stations.xls in a spreadsheet program and check the contents.

- Which file contains coordinates? Is there a way to link both files? How could we do that?

2.3   Import Spreadsheets

There are different ways in QGIS to import tabular data:

- Layer | Add Layer | Add delimited text layer. This is the standard importer that allows us to import delimited text files.
- Layer | Add Layer | Add spreadsheet layer. This tool can load spreadsheet files (*.ods, *.xls, *.xlsx) as a layer with options to use the first line as a header, ignore rows, and load geometry from x and y fields.

Here we'll use the second option, for which we need to install the *Spreadsheet Layers plugin*.

1. Start QGIS Desktop. Make sure you start a new project rather than continuing the previous one.

2. In the main menu, go to Plugins | Manage and Install Plugins and check if the *Spreadsheet Layers plugin* is installed (figure 2.2). If not, install it now.

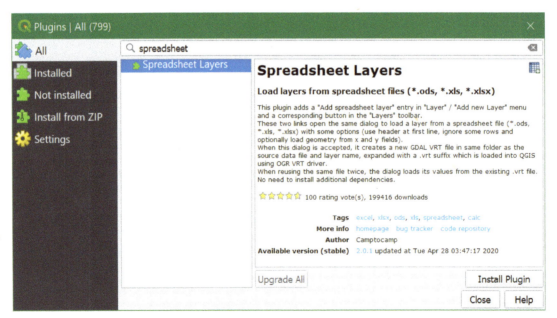

Figure 2.2: Install the Spreadsheet Layers Plugin

3. Now choose Layer | Add Layer | Add spreadsheet layer from the main menu (figure 2.3, on the next page).

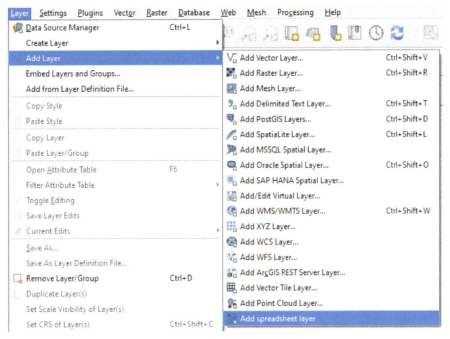

Figure 2.3: Add Spreadsheet Layer

4. In the dialog, browse to the file containing the locations of the meteorological stations (KNMI_stations.xls).

5. Fill in the dialog as in figure 2.4. Make sure to choose the correct *Geometry Fields* and *Reference system*. In addition, indicate the correct data types for the fields. For example, STN are station numbers and should be imported as *Integer*, while ALT(m) should be imported as *Real* numbers.

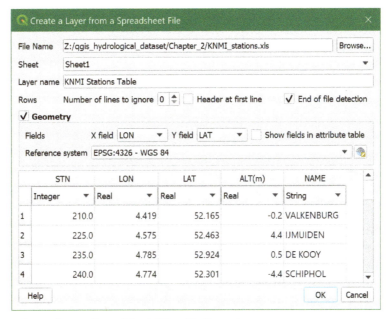

Figure 2.4: Create Layer from a Spreadsheet File with Geometry

Once you click OK, a map with the meteorological stations is displayed.

6. If you don't see the map, you probably need to zoom to the full extent of the map. Right-click on the layer name (*KNMI stations table*) and choose Zoom to layer.

7. Add the table with the temperature data in the same way. Because there is no geometry (coordinates) in the table, we should uncheck the box (figure 2.5).

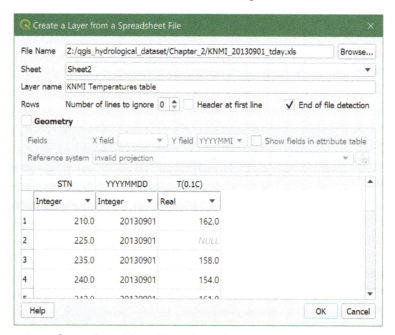

Figure 2.5: Create Layer from a Spreadsheet File without Geometry

> Spreadsheets or CSV files generally don't have projection information stored in the file. Therefore, it is important to find out which projection is used. When you see coordinates that look like decimal degrees, you can first try a Geographic Coordinate System (WGS 84, EPSG:4326) and check using an online backdrop layer from the QuickMapServices plugin to see if the points show up at the expected location. A good website for troubleshooting projection issues is https://ihatecoordinatesystems.com.

2.4 Convert Spreadsheet to Vector Layer

We now have the layer *KNMI stations table* saved as a temporary virtual layer. For further processing, we first need to convert it to a GIS vector format. In this case we'll convert it to a shapefile.

> In the previous chapter we have used the GeoPackage format, which has many advantages over the shapefile format. Shapefiles, however, are still very much used and therefore we also use them in this book. For an overview of limitations and advantages of these formats have a look at http://switchfromshapefile.org.

1. Right-click on *KNMI stations table* and choose Export | Save features as....

2. In the dialog, use the ... button to browse to the right folder to save the file as KNMI_ stations.shp. In order to change the projection to the local Dutch projection, choose *Amersfoort*

/ RD New for the *CRS* by clicking on the 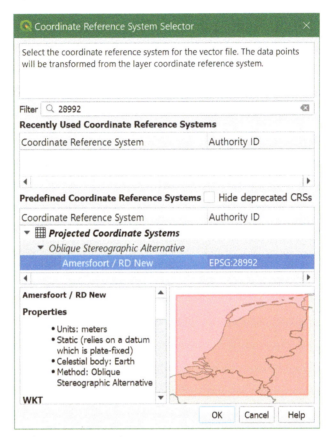 button. *Tip*: Use the *Filter* field to lookup EPSG code 28992 (figure 2.6).

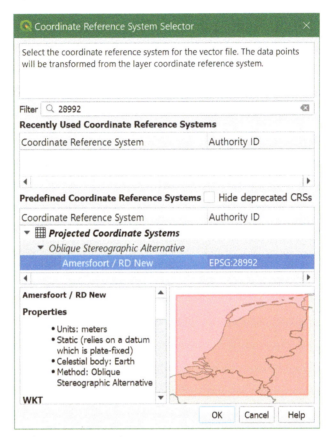

Figure 2.6: Coordinate Reference System Selector

> Here you see the advantage of using EPSG codes: they provide standard codes for each projection. It is useful to determine the EPSG code of the projection you want to use for your project.

3. Click *OK*. Now the dialog looks like figure 2.7, on the next page (also check the box *Add saved file to map* and make sure that *ESRI Shapefile* is chosen as the *Format*). You can see in the drop-down list that many formats are supported.

4. Click *OK* to proceed.

5. In the *Select Datum Transformations* pop-up, keep the default and click *OK*.

6. Remove the *KNMI Stations Table* from the layers list by right-clicking and selecting Remove Layer. Click *OK* to confirm. Be sure to remove the right one. If you hover your mouse over the layer item it will show the file name.

Note that with Remove Layer you only remove a layer from the layers list—the file will still be on your hard disk. If you want to delete the file from disk you need to go to the file in the *Browser* panel and right-click on the file name. Then choose Manage | Delete....

Although the knmi_stations.shp layer is in the EPSG:28992 projection (Amersfoort / RD New), the QGIS map canvas still uses the EPSG:4326 projection (lat/lon WGS 84) and has reprojected

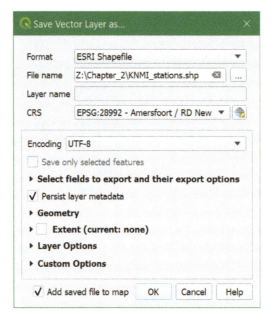

Figure 2.7: Save Vector Layer As Dialog

`knmi_stations.shp` *on-the-fly* for visualization. In order to visualize all layers in EPSG:28992 we have to change the QGIS project properties.

7. In the main menu choose `Project | Properties`.

8. Choose the *Coordinate Reference System (CRS)* tab (figure 2.8, on the facing page).

9. From the list of recently used coordinate reference systems, choose EPSG:28992 and click *OK*.

> Another way to set the project CRS to that of the layer, is to right-click on the layer and choose *Set CRS | Set Project CRS from Layer* from the context menu.

Note that the projection of the project is indicated in the lower right of the screen: 🌐 EPSG:28992 . You can always check there to see if the EPSG code is okay. This is also a button you can click to go to the `Project Properties | CRS` tab and change the on-the-fly reprojection.

2.5 Join Attribute Tables

The locations of the meteorological stations and the temperature data are still in separate tables. For further analysis, we need to combine them into one vector layer. In GIS terms, this is called a *join* operation. We can only join tables if they have a column in common.

1. Check the attribute table of *KNMI_stations* (right-click on KNMI_stations and choose `Open Attribute Table`) and in the same way, check the *KNMI Temperatures Table*.

   • Which column do both attribute tables have in common?

After determining which column both tables have in common we can join the data of *KNMI Temperatures Table* to the attributes of our shapefile `KNMI_stations.shp`.

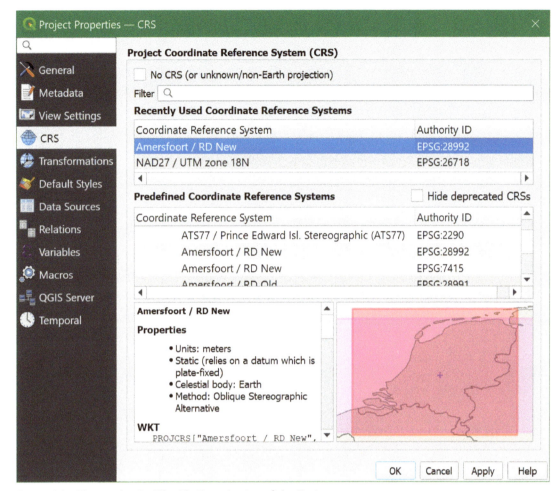

Figure 2.8: Change the On-The-Fly Reprojection of the Project

2. First close the attribute tables.

3. Next, right-click on *KNMI_stations* and choose `Properties`.

4. In the dialog choose the button *Joins*

5. Click the button and check if the dialog looks like figure 2.9, on the next page.

The common field is `STN` (the station number). We will join only the temperature field and give the column the prefix `Temp_`.

6. Click *OK*.

7. Click *OK* to perform the join operation.

Now check the attribute table of *KNMI_stations*.

- What happened?

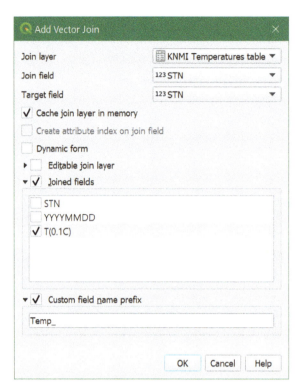

Figure 2.9: Add Vector Join

2.6 Edit the Joined Attribute Table

The joined attribute table needs two corrections: (1) features with missing data need to be removed, and (2) the temperatures have to be converted to the right units.

1. Click on row numbers with NULL or no values for temperature, while keeping the Ctrl key pressed.

Now the attribute table looks like figure 2.10, on the facing page.

2. In the attribute table click on ✏ above the table to toggle editing mode.

3. Click the 🗑 icon (in the toolbar above the attribute table) to remove the two features with missing data, then save the attribute table by clicking 📄.

The only problem now is that the temperatures in the table are in 0.1 °C. We need to convert the values to °C.

4. Click the *Open Field Calculator* 🧮 button.

5. In the *Field Calculator* dialog:

  • Verify that the *Create new field* box is checked. This will add a new field to the attribute table where the output of *Expression* (see below) will be stored.
  • At *Output field name* type T(C).
  • Choose Decimal number (real) for the *Output field type*.
  • *Output field length* is the number of digits, *Precision* is the number of decimal places. Set

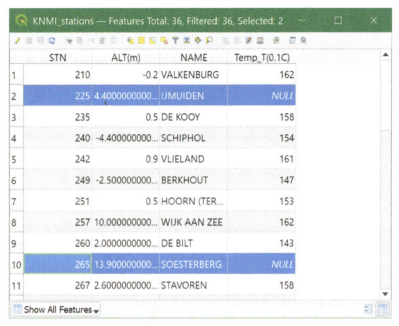

Figure 2.10: Select Features with Missing Data

it to 3 and 1 respectively.

- Under *Expression* fill in the dialog as shown in figure (2.11, on the next page). To avoid typos, the best practice is to double-click on the field name in the middle of the dialog screen and click the multiply button `*`. Then type 0.1 so the equation is "Temp_T(0.1C)" * 0.1.

6. Click *OK* to proceed.

Now check the result in the attribute table.

7. Click on the ✏ to toggle back to non-editing mode. Click *Save* to save the changes when asked and close the attribute table. If you made a mistake, don't save, but instead choose *Discard* to undo all changes since last save.

8. Now remove the table *KNMI Temperatures Table* from the layers list and check the attribute table of KNMI_stations.

- What columns do you see now?
- What can you conclude about the join function?

You could have saved the entire attribute table by saving *KNMI_stations* to a new shapefile using the previously used Export | Save Features As... function.

2.7 Interpolate Point Features to Raster

The final task is to interpolate the temperature values to a raster.

1. In the main, menu choose Raster | Analysis | Grid (Nearest Neighbor) (figure 2.12, on the following page).

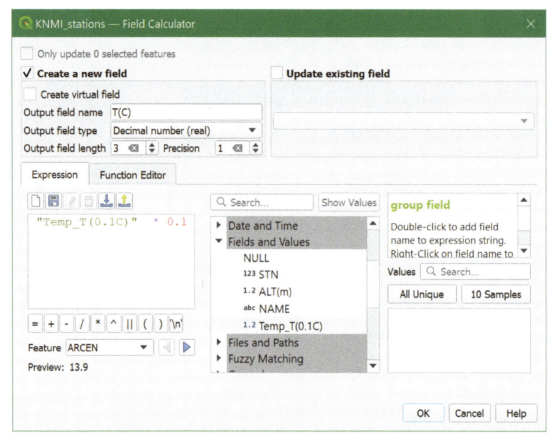

Figure 2.11: Create Temperature Field and Convert Units in the Field Calculator

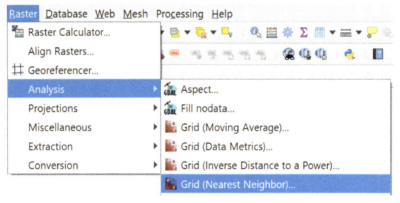

Figure 2.12: Interpolate to Raster using Nearest Neighbor Menu

2. In the dialog, specify the output file: tday_NN.tif by using the browse window and speci-
fying the .tif format.

3. In the *Advanced Parameters* section, select T(C) for *Z value from field*. This is the field that we
will interpolate to *Thiessen polygons*.

For the rest of the dialog, keep the defaults. The dialog should now look like figure 2.13, on
the next page.

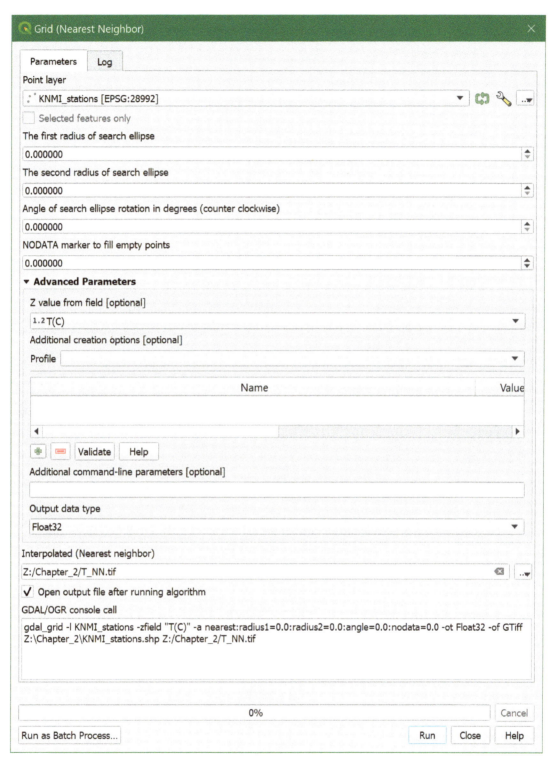

Figure 2.13: Interpolate to Raster using Nearest Neighbor Dialog

Note that the dialog generates a GDAL command. The tool is essentially a GUI for the gdal_grid command line tool (`https://gdal.org/programs/gdal_grid.html`).

4. Click *Run* to proceed.

5. Click *Close* to close the dialog.

6. Now repeat the interpolation using the Raster | Analysis | Grid (Inverse Distance to a Power) (IDW) algorithm. Name the resulting file tday_IDW.tif.

> With these interpolation tools from the main menu, you can't control the extent and spatial resolution of the output raster. In the Processing Toolbox there are interpolation tools where you can define the output extent and resolution—for example with the IDW Interpolation tool. The Processing Toolbox will be introduced in Chapter 3.

2.8 Styling the Results

To make sense of the interpolations, we need to style the raster layers.

1. Drag the *KNMI_stations* layer so that it is positioned above both the *T_NN* and *T_IDW* raster layers in the *Layers Panel*.

To provide some context, you will first add a basemap using the *QuickMapServices* plugin.

2. From the menu bar, choose Web | QuickMapServices | OSM | OSM Standard.

3. Open the *Layer Styling* panel by clicking the ✎ button. Set the target layer to *KNMI_stations*.

4. Select *Simple marker* component and set the *Fill color* to black and the *Size* to 2.8 mm.

5. Switch to the *Labels* tab ⒜ of the *Layer Styling* panel. Switch from *No Labels* to *Single labels*.

6. Set the *Label with* option to the NAME field.

7. To add more information to the map, you will now add the temperature to the label. Click the *Expression* ε button and use the *String Concatenation* ‖ operator and the *New line* 'n' operator to add the T(C) field to the label on a second line. We can add the degree symbol using the char() function, which returns a character associated with a *unicode code*. Your labeling expression should now read "NAME" || '\n' || "T(C)" || ' ' || char(0176) || 'C' (figure 2.14, on the facing page). Note that a space has been added between the value and the unit. Click *OK* to return to the *Layer Styling* panel.

The names are all in upper case. Let's change them to Title case.

8. Switch to the label *Formatting* tab and set the *Type case* to Title Case.

9. To center the label text, set the *Alignment* to *Center*.

10. Set the *Font* to Calibri with a *Style* of Bold and a *Size* of 10 points.

11. Switch to the *Label buffer* tab abc and check the *Draw text buffer* option. Set the *Opacity* to 50%.

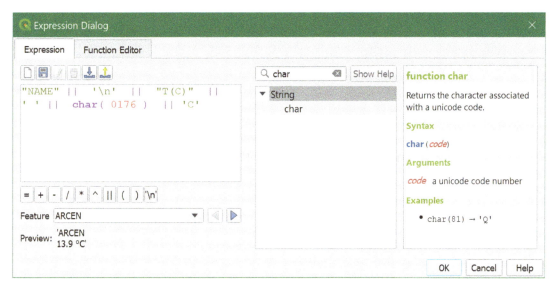

Figure 2.14: Expression for the Label of the Points with the Meteorological Stations

12. To give more separation between the labels and the feature icon, switch to the *Label placement* tab and set the *Distance* to 2 mm.

If labels are being cut off at the map boundary, review the previous chapter to remember how to use the *Automated placement settings* to prevent it.

Now you will turn your attention to the two interpolated rasters.

13. Start by making the *T_NN* layer the target layer in the *Layer Styling* panel.

14. Change the render from the default *Singleband gray* to *Singleband pseudocolor*.

For continuous rasters, we use the *Singleband pseudocolor* renderer. Although Thiessen polygons in the *T_NN* layer look discrete, the pixels have real numbers, which is not possible for discrete rasters.

15. For the *Color ramp*, choose *Spectral*. Click *Classify* if it doesn't automatically show the result.

By default, this color ramp is setup to make the lower values red and the larger numbers blue. This is counter-intuitive. These values represent temperature, so the higher values should be represented with warmer/red colors and the lower values with colder/blue colors.

16. To accomplish this, right-click on the color ramp and choose *Invert Color Ramp* from the context menu. In order to see the basemap, you will apply a **Blending mode**. In the *Layer rendering* section, choose *Multiply* for the *Blending mode*.

Your map should resemble figure 2.15, on the next page.

17. Next, turn off the *T_NN* layer and turn on the *T_IDW*. Use the *Layer Styling* panel to style the layer using the *Singleband pseudocolor* renderer with the same inverted color ramp. This time switch the *Mode* to *Quantile*. Also apply a Multiply *Blending mode* to this raster.

Your map should now resemble figure 2.16, on page 49.

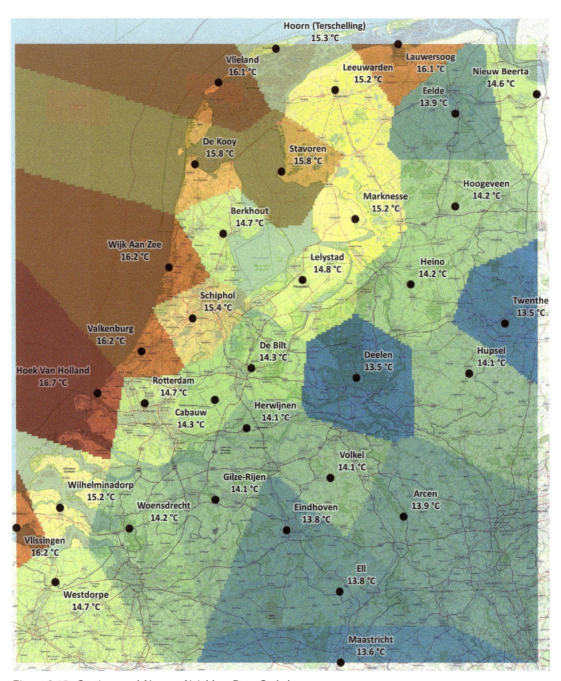

Figure 2.15: Stations and Nearest Neighbor Data Styled

- Which interpolation method is better? Why?
- Can you explain the temperature gradient in the map?

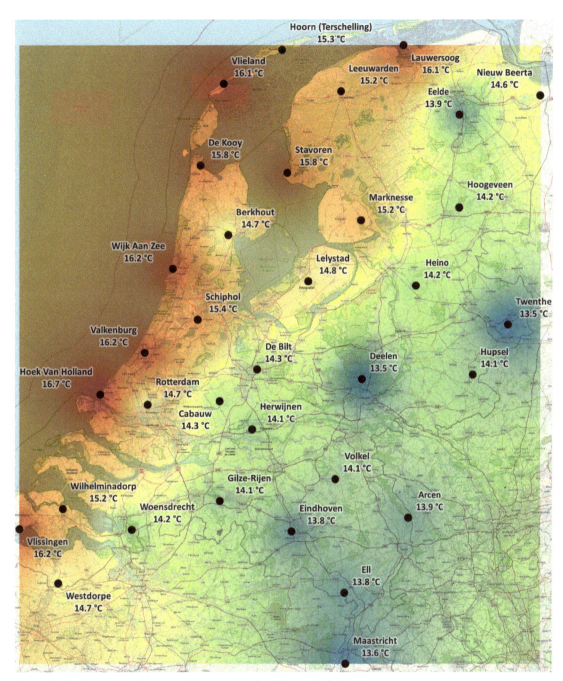

Figure 2.16: Stations and Inverse Distance Weighted Data Styled

3. Spatial Analysis with Map Algebra

3.1 Introduction

With *map algebra*, we can do calculations with raster layers. This is useful for spatial analysis. For example, when we need to evaluate different criteria to find suitable or unsuitable locations, we can use map algebra.

After this chapter you will be able to:

- use *raster attribute tables*
- apply *map algebra* for raster analysis
- distinguish *boolean*, *discrete*, and *continuous* rasters
- make legends for boolean, discrete, and continuous maps
- understand the use of *Nodata*
- use *logical operators*
- calculate *distances* from rasters
- *reclassify* rasters
- *convert* raster to vector points
- *sample* raster values with vector points

> For this chapter, the videos from the *QGISHydro Chapter 3* playlist at YouTube channel http://www.youtube.com/c/hansvanderkwast provide the theoretical background, results of the steps, and additional materials.

Case Description

The municipality of the (imaginary) oasis Aïn Kju Dzjis has hired you to analyze which wells are suitable for its inhabitants based on the following conditions:

Condition 1: The wells should be within 150 meters of houses or roads.

Condition 2: No industry, mine, or landfill within 300 meters of the wells.

Condition 3: The wells should be less than 40 meters deep.

You will use map algebra to perform the required analysis.

Figure 3.1, on the next page shows the workflow for this chapter.

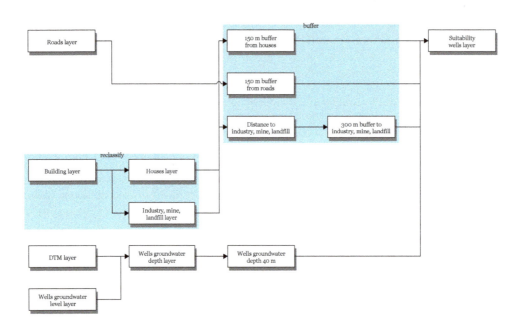

Figure 3.1: Workflow of Map Algebra for Finding the Suitability of Wells

3.2   Preparation

Checking the Metadata

For this task you are provided with the following raster layers: `buildg.tif`, `roads.tif`, `dtm.tif`, and `gwlevel.tif`. This data can be downloaded from `http://locatepress.com/qgis_hydrological`.

First we are going to add the exercise folder to the *Favorites* in the *Browser Panel*.

1. In the *Browser* panel, right-click on *Favorites* and choose `Add a directory` (figure 3.2). Alternatively, you can right-click on the folder and choose `Add as a Favorite`.

Figure 3.2: Add Favorites in the Browser Panel

2. Choose the folder where you have stored the layers.

3. Click on the little triangle to expand the contents of the folder.

4. Preview the maps and metadata of these raster layers by right-clicking on the layer and choosing *Layer Properties...* (figure 3.3, on the next page).

The *Layer Properties* window will open, showing the metadata of the layer (figure 3.4, on the facing page).

  • Check the metadata for all layers in this project. Note the file type, number of cells,

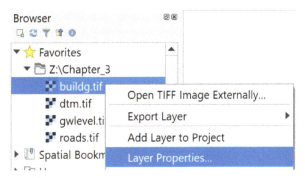

Figure 3.3: Layer Properties Menu

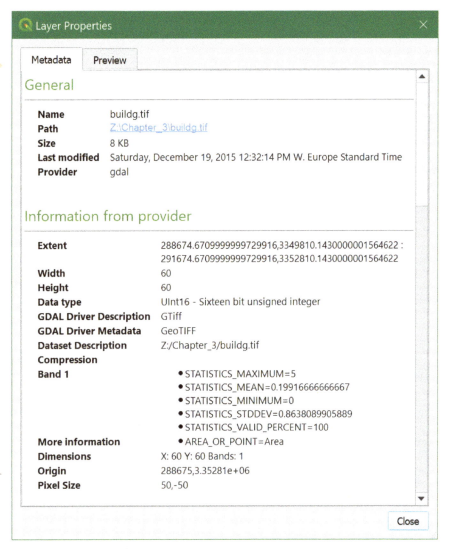

Figure 3.4: Layer Properties Window

projection, cell size, minimum value, and maximum value.

5. Select buildg.tif, roads.tif, dtm.tif and gwlevel.tif (keep the Ctrl button pressed and select the layers) and drag them to the map canvas.

6. Go to the *Layers* panel.

Adding Raster Attribute Tables

Raster layers, in general, have no attribute table. Each layer is one theme, represented by the values of the cells. The main raster data types are:

- Boolean rasters: cell values are zero or one, meaning True or False. An example is flooded versus non-flooded areas. They are styled using the *Paletted/Unique values* renderer.
- Discrete rasters: cell values are integers (whole numbers) representing classes. An example is a land cover map. They are also styled with the *Paletted/Unique values* renderer.
- Continuous rasters: cell values are real numbers (decimals) representing gradients in the landscape. The temperature maps calculated in Chapter 2 are examples of this. They are styled with the *Singleband pseudocolor* renderer.

In Chapter 4 we will cover a few more raster data types.

The GIS analyst has to determine the data type, because that is normally not stored in the raster layer. Knowledge of the data type is needed for applying the correct analysis tools and for presenting raster data with the correct styling.

QGIS now has the possibility to add *Raster Attribute Tables* or RAT's to the raster layers, which gives the opportunity to add multiple attribute fields to a single raster layer and choose a field for styling.

To use this functionality you need to install the *RasterAttributeTable* plugin.

1. Install the *RasterAttributeTable* plugin from the *Plugins Manager* (figure 3.5, on the next page).

We are going to add a RAT to the buildg layer. The first step is to apply a renderer to this discrete raster layer with land-use classes.

2. Select the buildg layer in the *Layers* panel and open the *Layer Styling* panel.

3. Choose the Paletted/Unique values renderer and click *Classify* to assign a random color to each cell value found in the raster layer (figure 3.6, on the facing page).

Without having a RAT, we could edit the labels of the cell values here to create a legend. In this case, we will do that in the RAT.

4. In the *Layers* panel, right-click on the buildg layer and choose New Attribute Table (figure 3.7, on the next page).

5. In the *New Raster Attribute Table* dialog (figure 3.8, on page 56), keep the default *GDAL auxiliary XML* as the format and click *OK*.

A message will indicate that the RAT is available.

6. Click *Open Raster Attribute Table* in the message or right-click on buildg in the *Layers* panel

Figure 3.5: Installing the RasterAttributeTable Plugin

Figure 3.6: The Paletted/Unique Values Renderer for Styling Boolean and Discrete Raster Data

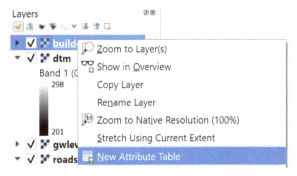

Figure 3.7: Create a New Raster Attribute Table

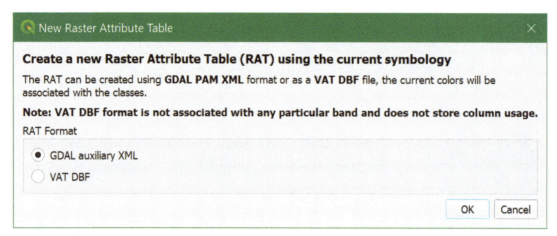

Figure 3.8: New Raster Attribute Table Dialog

and choose Open Attribute Table.

In the RAT, you will see several fields (figure 3.9):

- *RAT Color*: the colors assigned to the classes expressed in hexadecimal values
- *Value*: the cell values found in the raster layer
- *Count*: the number of cells in each class
- *Class*: the class label
- *RGBA*: Red, Green, Blue, and Alpha (opacity) values for the colors of each class

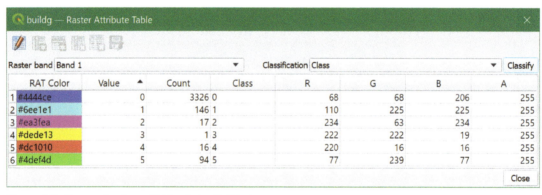

Figure 3.9: Raster Attribute Table of the Buildg Layer

Like with a vector attribute table, we can also add fields to RATs.

7. Click the *Edit Attribute Table* 🖊 button.

8. Click the *New Column* 🗔 button.

9. In the *Add new column* dialog, choose Class name as *Usage*, String as *Data type*, and type Land Use for the column *Name*. For *Insertion point*, choose *After* the Class column (figure 3.10, on the facing page). Click *OK*.

10. Type the land-use class names in the new *Land Use* field as in figure 3.11, on the next page.

11. Change the colors of the land-use categories to more intuitive colors. You can do this by clicking on a color in the *RAT color* field.

Figure 3.10: Add New Column to the RAT

12. Change the *Classification* field to Land Use and click *OK* in the pop-up to confirm that you want to replace the previous classification.

13. Save the edits by clicking the *Edit Attribute Table* 📝 button again and click *Yes* in the pop-up to confirm that you want to save the changes.

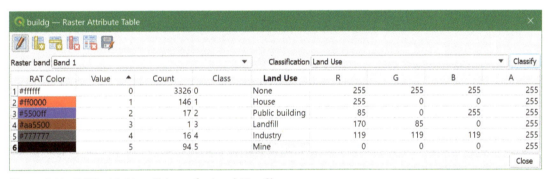

Figure 3.11: RAT with New Column for Land-Use Classes

14. Repeat the steps for the roads layer with the following classes:

- 0: no roads
- 1: dirt road
- 2: tarmac

The result should look like figure 3.12, on the following page.

> You can also create RATs from rasters styled with the Singleband Pseudocolor renderer. Then you need to change the Interpolation setting to Discrete so it uses class ranges.

Using the Processing Toolbox

The *Processing Toolbox* in QGIS provides a lot of tools for processing GIS data. Besides QGIS tools, it also has useful tools from other *processing providers* such as GDAL, GRASS, SAGA, and

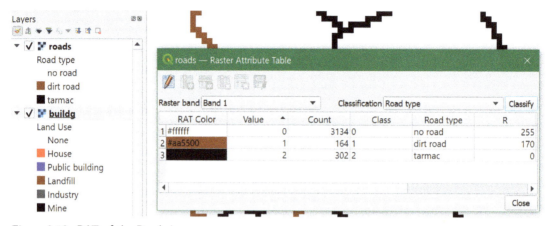

Figure 3.12: RAT of the Roads Layer

PCRaster.

1. First, activate the *Processing Toolbox* by choosing `Processing | Toolbox` from the main menu (figure 3.13).

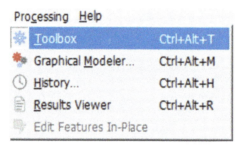

Figure 3.13: Open the Processing Toolbox from the Main Menu

> If you can't find the Processing Toolbox in the menu, then you need to activate the Processing plugin in the Plugins manager by checking the box. It's a core plugin that comes with QGIS.

3.3   Condition 1: Wells within 150 Meters of Houses or Roads

Let's first look at the houses. The houses are a class in the `buildg` layer.

Create a Boolean Layer with True for Houses and False for Other Buildings

If we want to create a Boolean layer with True (1) for houses and False (0) for the other classes in the `buildg` layer, we can use the *Raster Calculator*.

1. In the main menu, go to `Raster | Raster Calculator`.

2. In the *Raster Calculator* dialog, double-click on `buildg@1`, click the = button, and type `1` (figure 3.14, on the facing page).

Now the equation reads `buildg@1 = 1`, which means: if the `buildg@1` layer equals 1 (which is the houses class), then the output layer has True (value 1), else it has False (value 0). @1 means band 1. In our case we only use single band raster layers (multiple bands are used in other applications such as remote sensing).

Figure 3.14: Using the Raster Calculator to Create a Boolean Map With Houses

3. Call the output layer houses.tif and click *OK* to perform the calculation.

Following good practice, we are going to style this boolean layer.

4. For boolean layers, we also use the Paletted/Unique values renderer. Choose intuitive colors for True and False.

Create Zones of 150 Meters Around the Houses

Now that we have a layer with only houses, we can proceed with the calculation of the zones of 150 m around them. We're going to create a boolean layer which is True (1) within a zone of 150 m around houses and False (0) further than 150 m from houses.

1. In the main menu, choose Raster | Analysis | Proximity (Raster Distance) (figure 3.15, on the next page).

2. In the *Proximity (Raster Distance)* dialog, make sure the houses layer is selected as *Input layer*. Set the *Distance units* to Georeferenced coordinates. For *Maximum distance to be generated*, enter 150 meters and enter 1 for *Value to be applied to all pixels that are within the -maxdist of target pixels*. Set the *Output data type* to Byte (because we use only 0 and 1), and call the output *Proximity map* houses150m.tif. Leave the other settings at default (figure 3.16, on page 61).

Figure 3.15: Proximity Menu

3. Click *Run*, then *Close* when the calculation is done.

4. Style the boolean map. Make the True pixels green and the False pixels red (figure 3.17, on page 62).

Create Zones of 150 Meters Around the Roads

In a similar way, we can now calculate the 150 m buffer around all roads.

1. Repeat the steps used to calculate the zones of 150 m around the houses, but use the roads layer as the *Input layer* and name the output roads150m.tif.

2. Style the boolean map. Make the True pixels green and the False pixels red.

3.4   Condition 2: No Industry, Mine, or Landfill within 300 m from Wells

For the second condition, we first need to reclassify the buildg layer in such a way that the result is a boolean map with True (1) for industry, mine, and landfill and False (0) for the other classes. Then we need to calculate the distance to the True pixels, and finally calculate the pixels that are further than 300 m from industry, mine, and landfill.

Create a Boolean Raster with True for Industry, Mine, and Landfill, and False for Other Buildings

1. From the *Processing Toolbox* menu, choose Raster analysis | Reclassify by table (figure 3.18, on page 62).

The *Reclassify by table* dialog appears. We are going to reclassify the buildg raster using a *lookup table*.

2. Fill in the dialog exactly as shown in figure 3.19, on page 63.

Here you can identify a *nodata* value for the output layer for values that are excluded from the

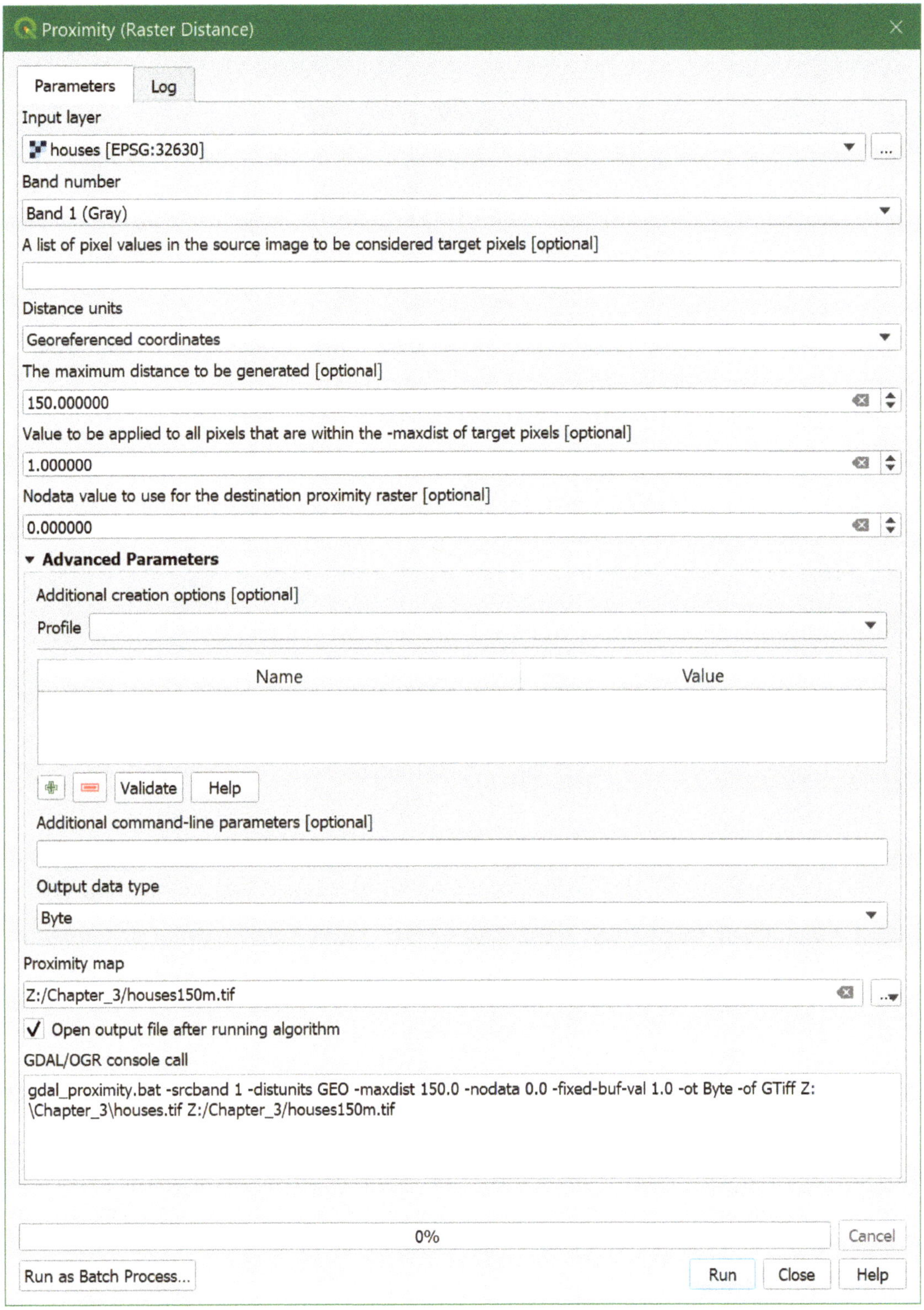

Figure 3.16: Proximity Dialog

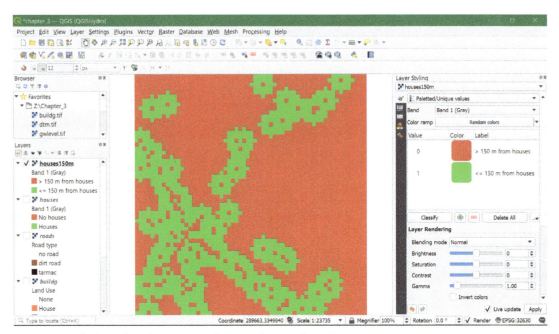

Figure 3.17: Styling of Boolean Map with Zone of 150 m from Houses

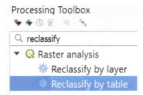

Figure 3.18: Reclassify by Table Processing Tool

lookup table. The *Range boundaries* define if values are included or excluded from the ranges in a row of the lookup table. Here we don't use ranges, but reclassify each value. Therefore, we choose min <= value <= max. For the *Output data type* we use Byte because the output values are whole numbers that fit in 8 bits (0 - 255).

3. Go to *Lookup table* and click `...`.

4. Fill in the lookup table as shown in figure 3.20, on page 64.

5. Click *OK* and *Run*. *Close* the dialog when the processing is finished.

6. Check the result: 1 for mines, industry, and landfills, 0 for the other classes. Use the *Identify* tool and click on the map. On the lower right panel you can find the identify results. It displays the value of the pixel of the selected layer in the *Layers* panel. You might have to resize the columns to see the pixel values.

 • Is industry a boolean, discrete, or continuous raster?

7. Style the industry layer.

Note that we could get the same result using the *Raster Calculator*.

 • Which expression would you have used for this in the *Raster Calculator*?

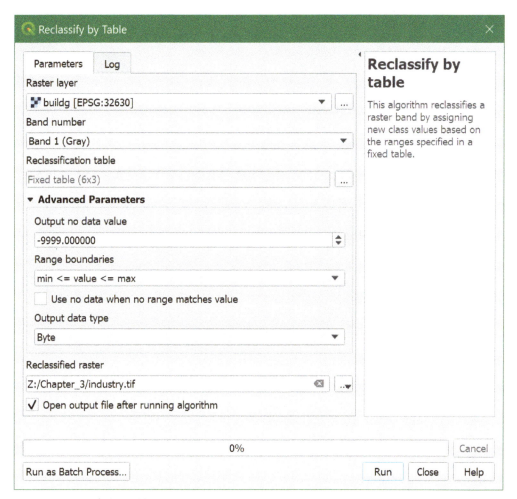

Figure 3.19: Reclassify by Table Dialog

Create Zones of 300 Meters Around Industry, Mine, and Landfill

With the boolean map with True for industry, mine, and landfill we can calculate the distance to these pixels. Because the *Proximity (Raster Distance)* tool does not allow us to assign values for pixels larger than a threshold, we have to calculate all distances and then use the *Raster Calculator* to calculate a boolean raster with True for pixels further than 300 meters from industry, mine, and landfill.

1. Open the *Proximity (Raster Distance)* tool.

2. Make sure that Industry is the *Input layer*. Keep all other settings as default. Name the *Output proximity map* inddist.tif (figure 3.21, on page 65).

3. Check the result.

  • Is the inddist layer a boolean, discrete, or continuous raster?

4. Make a legend that is appropriate for this raster type using *Singleband pseudocolor* as the render type. Use intuitive colors (e.g. a ramp from red to green).

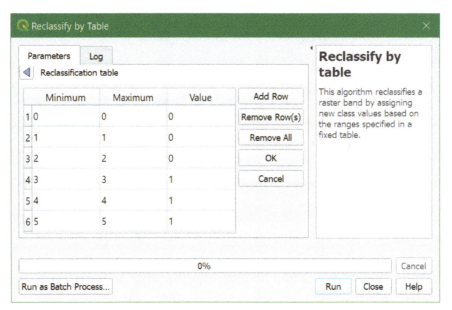

Figure 3.20: Lookup Table for Reclassification of Raster Values

5. Use the *Raster Calculator* to calculate `inddist@1 >= 300`. Call the output layer `ind300m.tif`

   • Is the `ind300m` layer a boolean, discrete, or continuous raster?

6. Style the `ind300m` layer (figure 3.22, on page 66).

3.5   Condition 3: Wells Less than 40 Meters Deep

For the last condition, we need to identify the wells that are less than 40 m deep. The `gwlevel` layer gives the absolute elevation of the groundwater level in the well in meters above sea level. In order to calculate the depth to the groundwater, we need to subtract this from the surface elevation given in the digital terrain model (DTM).

1. Open the *Raster Calculator*.

2. Subtract the absolute well depth from the DTM using this calculation: `dtm@1 - gwlevel@1`. Call the output layer `welldepth.tif`.

   • Is the `welldepth` layer boolean, discrete, or continuous?

3. Style the `welldepth` layer.

4. Next, calculate in the *Raster Calculator* a boolean map with wells less than 40 m deep. Call the output layer `notdeep.tif`.

5. Style the `notdeep` layer (figure 3.23, on page 66).

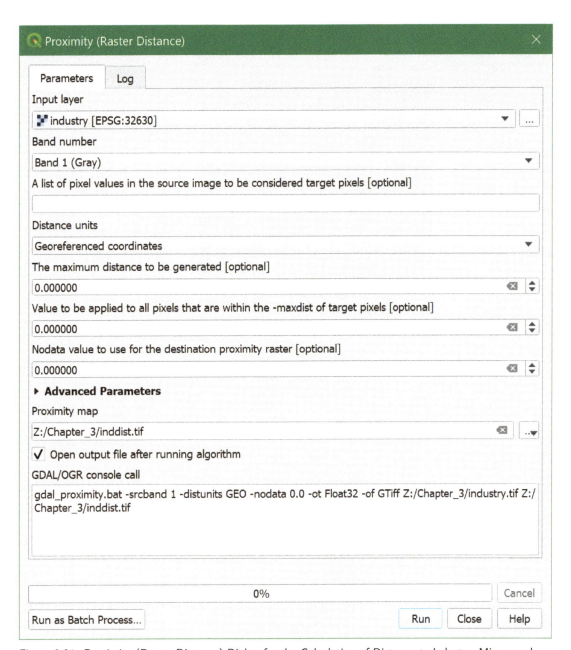

Figure 3.21: Proximity (Raster Distance) Dialog for the Calculation of Distance to Industry, Mines, and Landfill

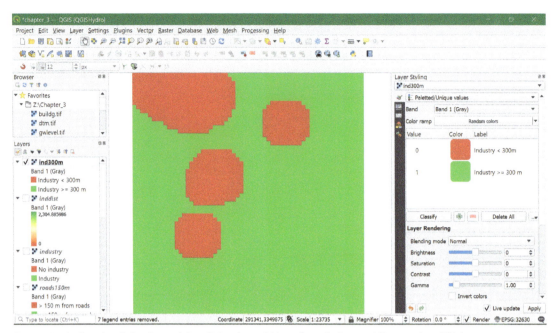

Figure 3.22: Result of the Second Condition (>= 300 m from Industry, Mines or Landfills)

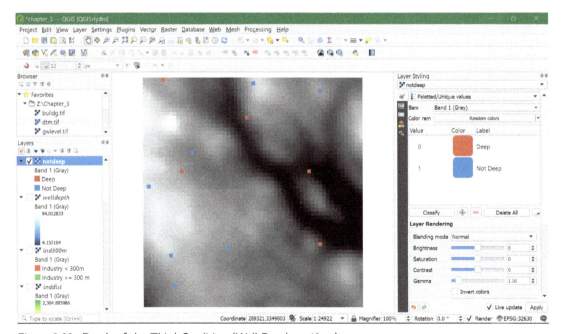

Figure 3.23: Result of the Third Condition (Well Depth < 40 m)

3.6 Combine the Three Conditions and Report the Results

Combine the Three Conditions

After calculating the boolean rasters for the three conditions, we need to combine them to come to the final result.

1. Use the *Raster Calculator* to combine the three conditions. Because all boolean results for the conditions need to be True we have to use the AND operator (figure 3.24). Call the output layer accessiblewells.tif.

Figure 3.24: Calculate the Combination of All Conditions in the Raster Calculator

2. Check the resulting raster layer and style the layer.

Convert Raster Cells to Point Vectors

To present the end result, we can style the accessiblewells layer. However, it is nicer to present the wells as point features on the map. Therefore, we need to convert the well pixels to point vectors.

1. In the *Processing Toolbox* go to Vector creation | Raster pixels to points (figure 3.25, on the next page).

2. In the dialog choose the accessiblewells layer as the input *Raster layer*. For *Field name*,

Figure 3.25: The Raster Pixels to Points Processing Tool

enter `Accessible`. The raster values will be saved in this field in the attribute table. Choose `wells.shp` as the output *Vector points* layer (figure 3.26).

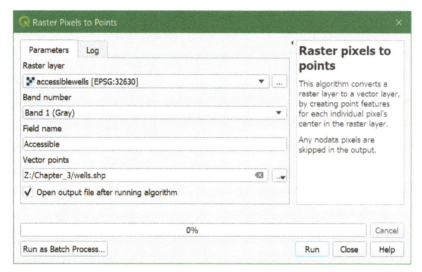

Figure 3.26: The Raster Pixels to Points Processing Tool Dialog

3. Click *Run*, then *Close* when the conversion is finished.

Sample Raster Values

The point vector layer `wells` only contains the field `Accessible`. It is however, more informative to also include other data in the attribute table. With the *Point sampling tool* plugin we can sample the raster layers in this project and add that information to the point attribute table.

1. Install the *Point sampling tool* plugin.

2. In the *Layers* panel, only check the boxes for the layers you want to sample and uncheck the others. Choose the following layers: `dtm`, `welldepth`, `gwlevel`, `notdeep`, `ind300m`, `roads150m`, and `houses150m`.

3. Click the *Point sampling tool* button 🐢.

4. In the *General* tab of the *Point Sampling Tool* dialog, choose `well` as the *Layer containing sampling points*, select all *Layers with fields/bands to get values from* and save the *Output point vector layer* to `wells_final.shp` (figure 3.27, on the next page).

5. Click the *Fields* tab.

6. Edit the *name* of the attribute that will be given to the output field if needed (figure 3.28, on the facing page).

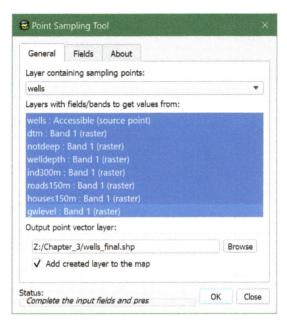

Figure 3.27: The General Tab of the Point Sampling Tool Dialog

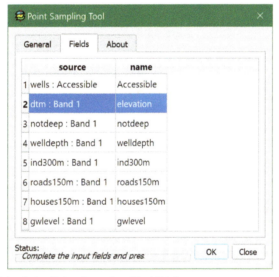

Figure 3.28: The Fields Tab of the Point Sampling Tool Dialog

7. Click *OK* and *Close*.

8. Open the attribute table of wells_final and check the result (figure 3.29, on the next page).

Style the Analysis Results

Now we can style the layers that we want to present. The first step in showing the results will be to style the wells_final data by whether the wells are accessible or not.

1. In the *Layers* panel, click on the 🖌 to open the *Layer styling* panel. Set wells_final as the target layer.

wells_final.shp — Features Total: 12, Filtered: 12, Selected: 0

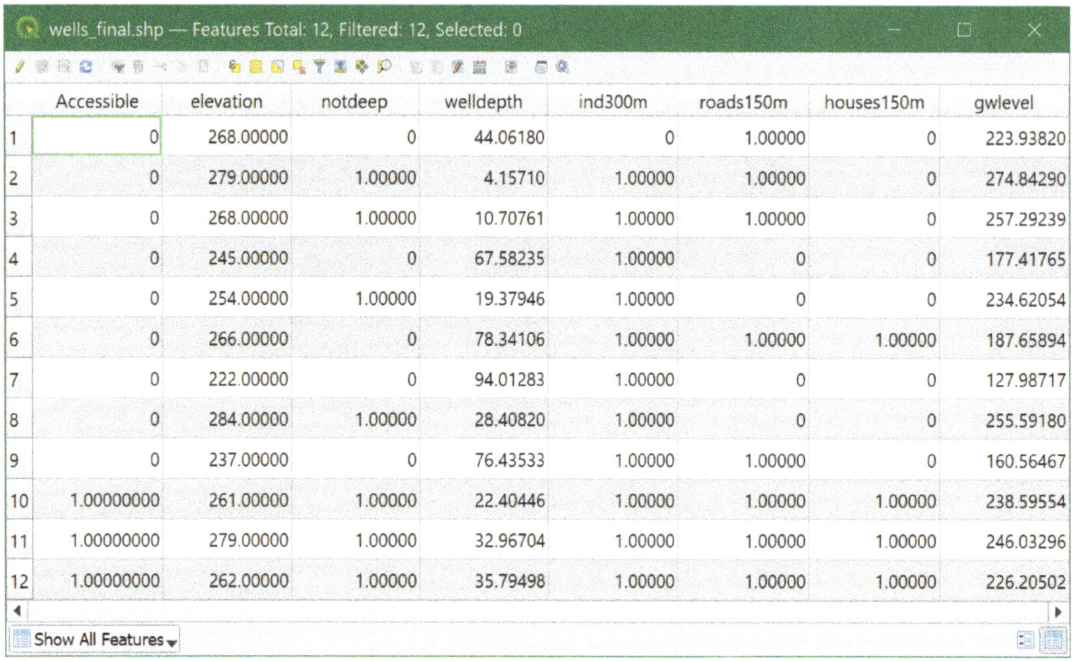

	Accessible	elevation	notdeep	welldepth	ind300m	roads150m	houses150m	gwlevel
1	0	268.00000	0	44.06180	0	1.00000	0	223.93820
2	0	279.00000	1.00000	4.15710	1.00000	1.00000	0	274.84290
3	0	268.00000	1.00000	10.70761	1.00000	1.00000	0	257.29239
4	0	245.00000	0	67.58235	1.00000	0	0	177.41765
5	0	254.00000	1.00000	19.37946	1.00000	0	0	234.62054
6	0	266.00000	0	78.34106	1.00000	1.00000	1.00000	187.65894
7	0	222.00000	0	94.01283	1.00000	0	0	127.98717
8	0	284.00000	1.00000	28.40820	1.00000	0	0	255.59180
9	0	237.00000	0	76.43533	1.00000	1.00000	0	160.56467
10	1.00000000	261.00000	1.00000	22.40446	1.00000	1.00000	1.00000	238.59554
11	1.00000000	279.00000	1.00000	32.96704	1.00000	1.00000	1.00000	246.03296
12	1.00000000	262.00000	1.00000	35.79498	1.00000	1.00000	1.00000	226.20502

Show All Features

Figure 3.29: The Final Attribute Table for the Wells

2. Change from a *Single symbol* to a *Categorized* renderer. Then set the *Value* to `Accessible`. Click *Classify*.

3. When using this renderer, QGIS will create a category to capture any NULL values. Here there are no wells with NULL values in the `Accessible` field. Select the entry where the *Value* reads *all other values* and click the *Delete* button to remove it. Give the wells with a value of 1 a green symbol with a size of 4 mm. Give the wells with a value of zero a red color with a size of 2 mm.

4. In the *Legend* column, rename the entry with a *Value* of 1 to *Accessible* and the entry with a *Value* of zero to *Inaccessible* (figure 3.30).

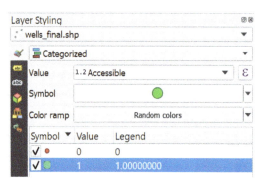

Figure 3.30: Styling the Final Wells

5. Scroll down in the *Layer styling* panel and find the *Layer Rendering* section. Expand it and select *Draw effects*. The *Customize effects* icon becomes active. Click on it to open the *Effects properties* panel.

Here you can add inner and outer glows, drop shadows, and other effects.

6. Click the box next to *Drop Shadow*. Select the *Drop Shadow* effect so that you are seeing the parameters of the *Drop Shadow*. Reduce the *Offset* distance to 1.0 (figure 3.31). Click the *Go back* ◀ button to return to the main styling panel.

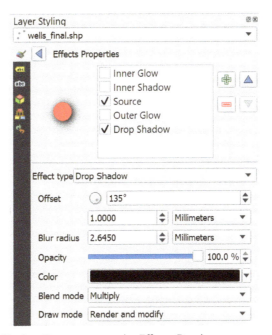

Figure 3.31: Setting Drop Shadow Parameters on the Effects Panel

7. Switch to the *Labels* tab ⟨abc⟩ of the *Layer styling* panel.

8. Change the setting from *No labels* to *Single labels*. Set the *Value* field to welldepth.

9. You will use a labeling expression very similar to that used at the end of the Preparing Data from Hard Copy Maps chapter, on page 29. You will set the expression up so the labels read something like: *Well Depth: 76.4 m.*

10. Click the *Expression* ε button to open the *Expression Dialog* window.

11. First, add the string *Well Depth* in front of the well depth value by adding 'Well Depth: ' before the existing expression.

12. To accomplish the second component, you will use the format_number function. Use the search box to find the format_number function. Insert it right before the welldepth field. Remember, this function requires a number and a number of decimal places. The number will be the welldepth field and the number of places 2.

13. Use what you learned about the *String Concatenation* ⟨ || ⟩ operator and the *New line* ⟨'\n'⟩ operator to nicely format the label.

14. Your expression should resemble 'Well Depth:'|| '\n' || format_number(welldepth,2) || ' m'. Click *OK* to return.

15. Set the *Font* to a sans serif font with a *Size* of 9 points.

16. Switch to the *Placement* 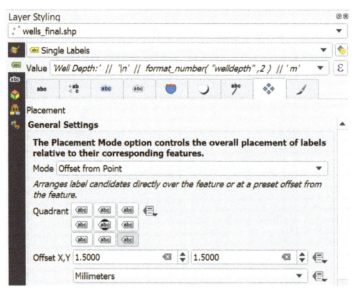 tab and set the *Mode* to *Offset from point*. Select the lower right *Quadrant* placement with an *Offset X Y* setting of 1.5 mm each (figure 3.32).

Figure 3.32: Label Placement Settings

17. Finally, click the *Automated placement settings* button to open the *Automated Placement Engine* window. Uncheck the box for *Allow truncated labels on edges of map* option. This will prevent labels from being cut off.

To complete the styling, you will work with the dtm layer. Make sure that all layers are switched off except wells_final on top and dtm below.

18. Make the dtm layer the target layer in the *Layer styling* panel.

19. Choose the Singleband pseudocolor renderer.

20. Click the drop-down for the *Color ramp* and choose *Create new color ramp*.

21. The *Color ramp type* window opens. Choose *Catalog:cpt-city* as the type. Click *OK*.

22. The *Cpt-city Color Ramp* window will open. Select the *Topography* category.

23. Choose *sd-a* and click *OK*.

> Did you know that you can save this color ramp to your style library? To do this, access the color ramp context menu and select *Save color ramp*. The *Save New Color Ramp* window will open allowing you to give it a *Name* and provide *Tag(s)*.

24. Scroll down to the *Layer Rendering* section and set the *Blending mode* to Multiply.

25. Finally, right-click on the dtm layer and choose *Duplicate* from the context menu.

26. Rename dtm copy to hillshade (right-click on the layer and choose Rename) and turn it on.

27. Make the `hillshade` layer the target layer in the *Layer styling* panel. Change the renderer to *Hillshade*. Scroll down to the *Layer Rendering* section and set the *Blending mode* back to *Normal* (figure 3.33).

28. In the *Resampling* section, set *Zoomed in* to `Bilinear` to create a smooth visualization. You can also do this for the `dtm` layer.

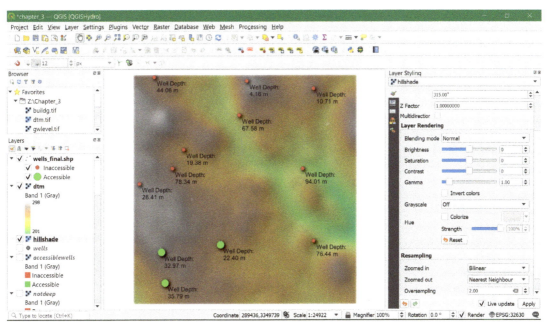

Figure 3.33: Final Analysis Results Styled

4. Stream and Catchment Delineation

4.1 Introduction

One of the most important uses of GIS in hydrology is the delineation of streams and catchments. This chapter presents a generic workflow for stream and catchment delineation for areas where only open data is available. The Shuttle Radar Topography Mission (SRTM) 1 Arc Second DEM will be used. The workflow will be applied to the Rur catchment in Germany.

After this chapter you will be able to:

- *download SRTM DEM tiles* using the SRTM-Downloader plugin
- *mosaic raster layers* into a virtual raster
- *reproject* rasters
- *create subsets* of rasters
- *fill sinks* in a DEM
- calculate *flow direction* raster layers
- style flow direction raster layers using circular color ramps and arrows in 2D and 3D
- calculate *Strahler orders*
- *delineate streams*
- *delineate catchments*
- style layers for visualizing catchments
- store project and data in a GeoPackage

Generic Workflow for Catchment Delineation

In order to delineate a catchment from a DEM in GIS we need to follow these steps:

1. Download the DEM tiles of your study area. Make sure that the tiles cover at least the study area and that the catchment you want to derive is covered completely. Better to take it a bit larger to avoid boundary effects.

2. If your study area is covered by multiple DEM tiles, you need to mosaic (merge) the tiles to create a single raster DEM layer.

3. The DEM tiles might be in a different coordinate system than desired. In that case you have to reproject the DEM layer to the projection you want to use in the study area.

4. If the merged DEM is much larger than your study area, you can subset (clip) it to a smaller area to reduce calculation time.

5. Interpolate voids when pixels with no data exist in the DEM.

6. Make a hydrologically correct DEM by filling sinks and removing spikes.

7. Calculate the flow direction for each cell.

8. Derive the drainage network.

9. Calculate the catchment for the outflow point of the catchment.

When a map with the stream network already exists, the procedure can be improved by "burning" the river network into the DEM. In that way the DEM is always lower at rivers and runoff will follow the actual river network. This is beyond the scope of this chapter.

Figure 4.1 shows the workflow.

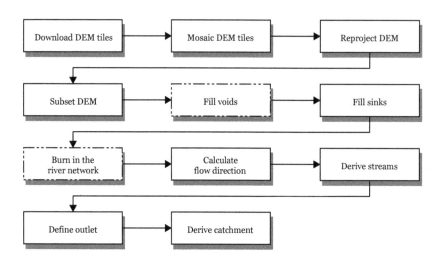

Figure 4.1: Workflow for Stream and Catchment Delineation

Data for this chapter can be downloaded from `http://locatepress.com/qgis_hydrological`. The next section covers how to download the SRTM tiles using the *SRTM-Downloader* plugin. These can also be downloaded from the course data repository.

> For this chapter, the videos in the *QGISHydro Chapter 4* playlist at YouTube channel `http://www.youtube.com/c/hansvanderkwast` provide the theoretical background, results of the steps and additional materials.

4.2 Download DEM Tiles

For the Rur study area, we will download the tiles from the SRTM 1 Arc-Second global data set. Since the end of 2014, a 1-arc second global digital elevation model (approximately 30 meters at the equator) has been released as open data. Most parts of the world have been covered by this dataset ranging from 54 degrees south to 60 degrees north latitude. A description of the SRTM data products can be found here: `http://loc8.cc/srtm1arc`.

The following steps explain how to download the SRTM DEM tiles for the study area using the *SRTM-Downloader* plugin.

1. Start QGIS Desktop. We'll start with a new project. In the browser panel, add the folder with the data for this chapter to the *Favorites* as you learned in the previous chapter.

2. Drag the `boundingbox.shp` layer from the *Browser* panel to the map canvas.

This polygon defines the boundary of our initial analysis. Normally you have to create such a polygon yourself or use the map canvas coordinates. We need this to clip the DEM to the study area. Note that the projection of the project is now set to the projection of the boundingbox layer.

3. Install the SRTM-Downloader plugin from the *Plugins Manager* (figure 4.2).

Figure 4.2: Installing the SRTM-Downloader Plugin

4. Click the *SRTM-Downloader* button in the toolbar.

5. In the *SRTM-Downloader* dialog, click *Set canvas extent*. It will fill in the coordinates of your map canvas in GCS coordinates.

6. Keep the other settings as default and click the *Download* button.

7. The first time you use this tool, you will see a pop-up that asks for your login. Follow the link to make an account and fill in your credentials in the pop-up. Click the box to *Save Credentials* and click *OK* to proceed with the download.

Six SRTM tiles are now downloaded in the SRTM HGT raster format and added to the map canvas (figure 4.3, on the next page).

8. Click *OK* when the pop-up appears that the download is complete and click *Close* to close the *SRTM-Downloader* dialog.

9. Move the boundingbox layer to the top so you can see which area of the DEM tiles falls within the study area.

> Alternatively, you can download the SRTM tiles from USGS Earth Explorer (http://earthexplorer.usgs.gov) using the same credentials. You can also download many other free datasets from the site. Another great plugin is the *OpenTopography DEM Downloader* plugin. The latter can be used to download different DEM products clipped to a specified extent.

10. Right-click on one of the tiles and choose Properties to review the metadata (figure 4.4, on page 79).

- What is the projection?
- What are the map units?

Figure 4.3: Download SRTM tiles with the SRTM-Downloader Plugin

4.3 Mosaic DEM Tiles

Before we proceed, we have to merge the DEM tiles, which in GIS terminology is called mosaic.
Here are two ways to mosaic the tiles:

- Merge the tiles into one physical file (e.g. GeoTIFF).
- Merge the tiles into a virtual file (.vrt).

The first option is slower. If we have many tiles, we prefer to create a file that virtually merges
all the tiles. That will be done with the following steps:

1. In the main menu, choose Raster | Miscellaneous | Build Virtual Raster (figure 4.5, on
the next page)

2. In the *Build Virtual Raster* dialog, you can choose each file individually or merge all files in
a directory. We can also merge the files that are visible in the map canvas. We'll use the last
option (figure 4.6, on page 80):

- At *Input layers* click ... .
- Use the *Select all* button to select the four tiles and click *OK* (figure 4.6, on page 80).
- Browse to the location where you want to save the output file and give it the name dem_
mosaic.vrt.
- By default, *Resolution* is set to average. In our case, the files all have the same resolution
(1 Arc Second).
- Make sure the box before *Place each input file into a separate band* is unchecked. This needs
to be checked only if you want to create a mapstack, i.e. with remote sensing bands.

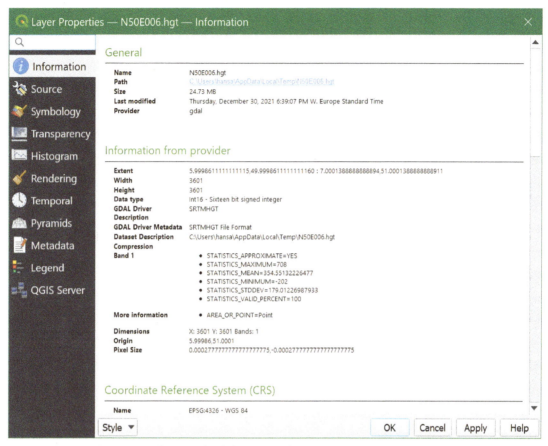

Figure 4.4: Layer Properties of a Downloaded SRTM Tile

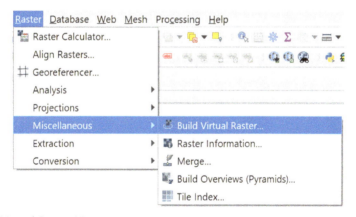

Figure 4.5: Build Virtual Raster Menu

The dialog should now look like that shown in figure 4.7, on the following page.

3. Click *Run* to run the algorithm. Click *Close* to get back to the main screen where you can see the merged DEM.

You'll notice that in the map canvas, the borders of the tiles are not visible anymore in the merged DEM because QGIS stretches the greyscale using the minimum and maximum cell values of the entire merged DEM. This is only for visualization, the values in the tiles are the

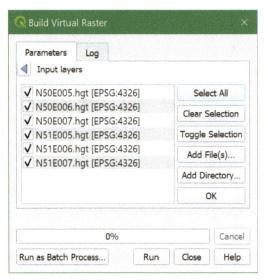

Figure 4.6: Multiple Selection Dialog

Figure 4.7: Build Virtual Raster Dialog

same as in the mosaic.

4. Now remove the individual tiles (not the dem_mosaic) from the layers list by selecting them while the Ctrl button is pressed. Then right-click one of the tile names and select Remove Layer. Click *OK* to confirm. This will remove the tiles from the screen, but not from the hard disk.

4.4 Reproject DEM

The DEM is in its original Lat/Lon Geographic Coordinate System (GCS) with datum WGS 84 (EPSG: 4326). It is not recommended to use a GCS for DEM analysis, because the Z units (e.g. meters) are different than the X and Y units (degrees). We need to choose a projection for our project. If the project covers one country, we can choose a national projection. In our case, however, the project covers multiple countries (Germany, the Netherlands, and Belgium). Therefore, we will use a global projection: UTM Zone 32 North, with WGS 84 as datum.

We can find the EPSG codes at `http://www.spatialreference.org`.

1. Use the website to search for `UTM 32N wgs 84`. You can leave QGIS running and open a browser.

Figure 4.8 shows the result.

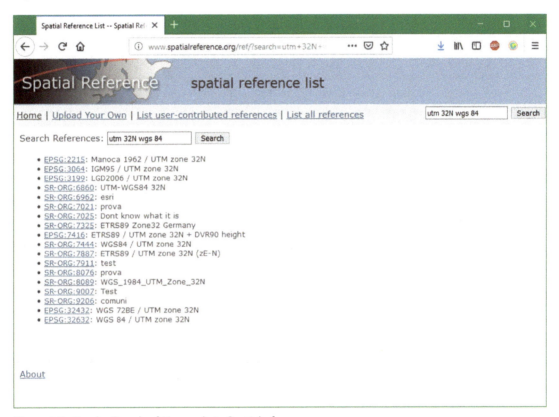

Figure 4.8: Results Search of Keywords in Spatialreference.org

We need to take a look at the EPSG codes. We will use EPSG: 32632 throughout this project—clicking on it will provide more details. Note that this is the same projection the `boundingbox` layer has and was used for setting the projection of our project when we loaded the `boundingbox` layer as our first layer.

Now we are going to reproject the DEM from unprojected GCS (Lat/Lon WGS 84 - EPSG: 4326) to UTM Zone 32 North / WGS 84 (EPSG: 32632).

2. From the main menu, choose `Raster | Projections | Warp (Reproject)` (figure 4.9, on the following page).

3. In the *Warp (reproject)* window, choose the `Project CRS: EPSG:32632 - WGS 84/UTM zone 32N` as *Target CRS*.

4. Now complete the dialog:

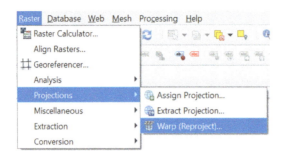

Figure 4.9: Reproject Rasters using Warp Menu

- For the *Resampling* method, choose Nearest Neighbor to preserve the elevation values of the original files.
- Set the *No data value for output bands* to -9999 because when the raster layer is reprojected there will be "no data" at the borders. In this way, we define that "no data" has a value -9999 and will not be visualized and treated as "data".
- Set the *Output file resolution* to 30 m.
- Browse to your exercise folder and name the output file dem_reprojected.tif. Make sure you choose a GeoTIFF as output format.

The dialog should now look like that shown in figure 4.10, on the next page.

> Note the gdalwarp command that will be executed in the background. Under *Advanced Parameters*, we could have specified the extent of the output file based on the boundingbox layer. In that way the dem_mosaic layer is reprojected and clipped in one step.

5. Click *Run* to run the algorithm. After running, click *Close* to close the window.

The reprojected DEM now appears in the map canvas.

6. You can now remove the dem_mosaic layer.

This is a good time to save your project.

4.5   Create a Subset of the DEM

In order to reduce the calculation time of the algorithms, we will subset (or clip) the raster layer to the boundingbox polygon.

1. From the main menu, select Raster | Extraction | Clip Raster by Mask Layer (figure 4.11, on the facing page).

2. In the *Clip Raster by Mask Layer* dialog, choose dem_reprojected for the *Input layer*. For *Mask Layer*, choose boundingbox. Check the box to *Keep resolution of input raster* and keep the defaults for the other options. Choose dem_subset for the *Clipped (mask) output* and click *Run*. Click *Close* when done (figure 4.12, on page 84).

Figure 4.10: Warp (Reproject) Dialog

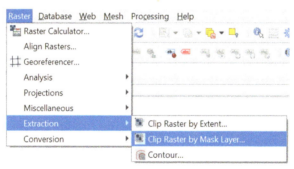

Figure 4.11: Clip Raster by Mask Layer Menu

Often you won't have a bounding box shapefile. In that case you can choose `Raster | Extraction | Clip Raster by Extent`. Then you can drag a box in the map canvas with a backdrop layer (e.g. OpenStreetMap) and use that for clipping. While using that, make sure your on-the-fly reprojection is similar to the CRS of the layer that you want to clip, because the map canvas coordinates are used by the algorithm.

3. Now you can remove `dem_reprojected` from the layers list as we have done before for other layers that are no longer needed.

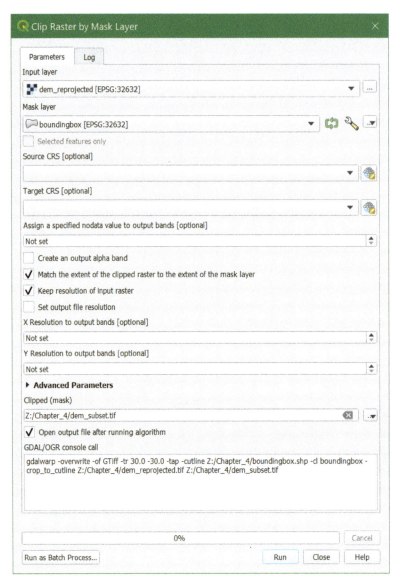

Figure 4.12: Clip Raster by Mask Layer Dialog

Styling the DEM

By default, QGIS styles raster layers with the Singleband gray renderer. As you have learned in Chapter 3, the software does not know the raster data type (e.g. boolean, discrete or continuous) and how the user wants to present the information. To help interpret the results, it is good practice to intuitively style your layers.

The DEM is a continuous raster. Continuous rasters represent gradients and can therefore contain real numbers (also called decimal numbers or floating point). Continuous rasters are styled in QGIS using ramps from the Singleband pseudocolor renderer in the *Layer Styling* panel.

4. Select the dem_subset layer and click 🖌 (or press F7).

5. In the *Layer Styling* panel, choose Singleband pseudocolor from the drop-down list.

6. Click the arrow at *Color ramp* and choose `Create New Color Ramp` from the drop-down menu (figure 4.13).

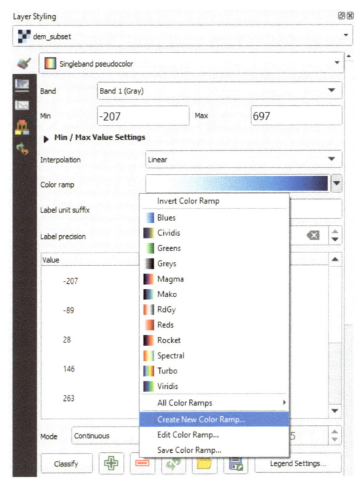

Figure 4.13: Create New Color Ramp

7. In the pop-up *Color ramp type* dialog choose `Catalog: cpt-city` from the drop-down list (figure 4.14).

Figure 4.14: Color Ramp Type Dialog

The `cpt-city` catalog has a lot of useful preset color ramps.

8. Choose `Topography | Elevation`. Note that *cd-a* and *sd-a* are also nice choices (figure 4.15, on the following page).

9. Click *OK* to close the dialog.

This gives us more intuitive colors in the DEM where we can clearly distinguish higher and lower areas (if you don't see the colors applied, click *Classify*).

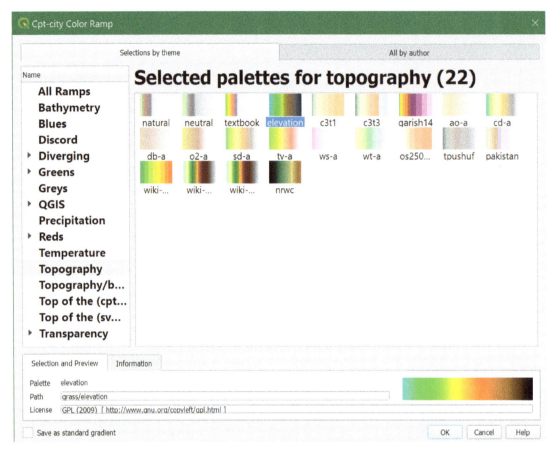

Figure 4.15: Cpt-city Dialog

Now we will further improve the visualization.

10. Right-click on the dem_subset layer and select Duplicate Layer (figure 4.16).

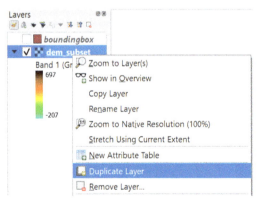

Figure 4.16: Duplicate Layer Menu

This creates a copy of the dem_subset layer called dem_subset copy.

11. Uncheck the dem_subset layer and rename the dem_subset copy layer to hillshade.

12. In the *Layer Styling* panel, which should still be open, make sure that the hillshade layer

is now selected. In the drop-down list change *Singleband pseudocolor* to *Hillshade*.

Now the hillshade layer is visualized with a shading.

- Which direction is the illumination coming from?
- Is this possible in reality?

> Hillshade gives the best results with an artificial illumination in the northwest, which in reality cannot exist in the Northern Hemisphere. If you move the dial in the *Layer Styling* panel to the southwest, you will see an inverted relief. Also note that there is a *Resampling* section. The default resampling method for both *Zoomed in* and *Zoomed out* is *Nearest Neighbor*. This method is fine for categorical data however, elevation is considered continuous data. You should therefore choose a *Zoomed in* resampling method of *Bilinear* and a *Zoomed out* resampling method of *Cubic*.

Next, we're going to blend the DEM with the hillshade layer.

13. Switch on the dem_subset layer by checking the box.

14. In the *Layer Styling* panel make sure the dem_subset layer is selected. In the *Layer Rendering* block of the panel, change the *Blending mode* to *Multiply* (figure 4.17).

Figure 4.17: Change the Blending Mode

As you can see, blending gives a much nicer effect than transparency. With transparency the colors will fade. Now we can clearly see the elevation differences—the gradient from south to north and the valleys where we expect the streams (see figure 4.18, on the next page).

> Blending modes allow for more elegant rendering between GIS layers. They can be much more powerful than simply adjusting layer opacity. Blending modes allow for effects where the full intensity of an underlying layer is still visible through the layer above. There are thirteen blending modes available. More information is available in the QGIS documentation: https://edu.nl/rcddd.

4.6 Fill Sinks and Calculate Flow Direction

Raw, unprocessed DEMs have artifacts such as depressions. Artifacts are a result of the DEM acquisition process and must be removed before a DEM can be used for hydrological analysis, like catchment and stream delineation or hydrological modeling. There are several algorithms

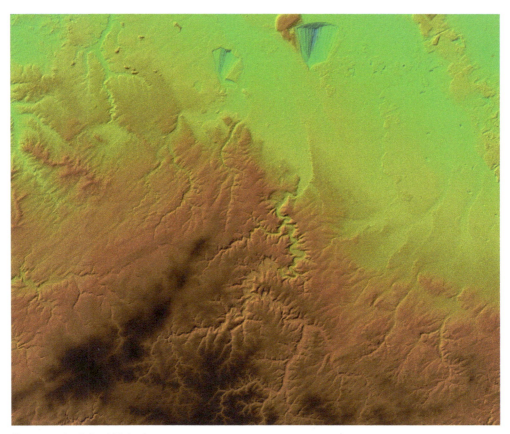

Figure 4.18: Color Hillshade

for filling sinks. We will use the *lddcreate* tool from the *PCRaster Tools* plugin. This tool fills the sinks and creates a flow direction map (also called local drain direction map) at the same time. The resulting flow direction map can be used in the next steps.

First, we will install the *PCRaster Tools* plugin. Then we will convert the DEM GeoTIFF to the PCRaster format (`.map`, which is a GDAL supported raster format). Next we will fill the sinks and derive the flow direction with the *lddcreate* tool. Finally, we will style the flow direction map.

Install the PCRaster Tools plugin

The *PCRaster Tools* plugin adds ~100 tools to the *Processing Toolbox*. These are tools from the PCRaster Python package. These tools are useful for map algebra and environmental modeling in the fields of geography, hydrology, and ecology. More information about PCRaster can be found at `http://www.pcraster.eu`.

Before you can install the *PCRaster Tools* plugin you need to install PCRaster.

On Windows, you can install PCRaster with the OSGeo4W installer. Alternatively, you can install PCRaster and QGIS with Conda (`https://docs.conda.io`).

In this section, we will explain how to install PCRaster Tools on Windows using the OSGeo4W installer. For installation with Conda, please check the plugin documentation here: `https://jvdkwast.github.io/qgis-processing-pcraster`.

1. Save your QGIS project and close QGIS.

2. Run the *OSGeo4W Setup,* which comes with your QGIS version. You can find it in the Windows Start Menu (figure 4.19).

Figure 4.19: OSGeo4W Setup in the Windows Start Menu

3. In the *OSGeo4W Setup* wizard, choose *Advanced Install,* click *Next.*

4. Choose *Install from Internet* and click *Next.*

5. Select the *root install directory* or keep the defaults; click *Next.*

6. Select the *local package directory* or keep the defaults; click *Next.*

7. Select your internet connection; click *Next.*

8. Choose one of the download sites; click *Next.*

9. In the *Select Packages* window, search for `pcraster`.

10. Click the arrows icon to change from *skip* to a PCRaster version to install (figure 4.20).

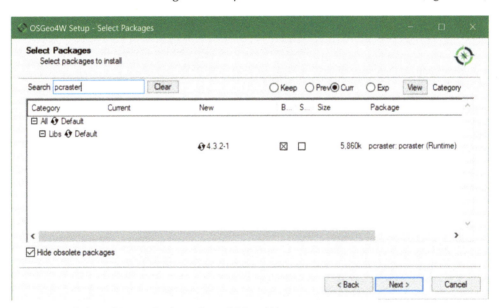

Figure 4.20: Install the PCRaster Package from OSGeo4W

11. Click *Next* to run the installation.

12. Click *Finished* to close the *OSGeo4W Setup* wizard.

13. Start QGIS Desktop.

14. In the main menu, choose `Plugins | Manage and install plugins....`

15. In the dialog, search for `pcraster` and install the *PCRaster Tools* plugin (figure 4.21, on the following page).

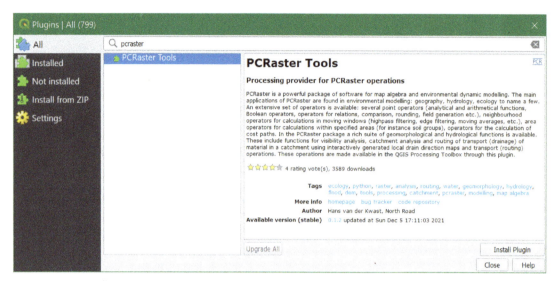

Figure 4.21: Installing the PCRaster Tools Plugin

After successful installation, you will see the *PCRaster Tools* in the *Processing Toolbox* (figure 4.22).

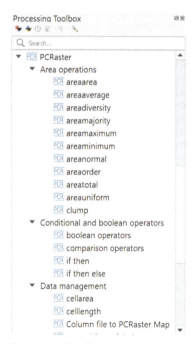

Figure 4.22: PCRaster Tools in the Processing Toolbox

Convert GeoTIFF to PCRaster Format

In order to use the *PCRaster Tools*, we need to convert rasters to the PCRaster format, which is a GDAL supported format. This is necessary because PCRaster is strict with raster data types.

Previously, you learned about boolean, discrete, and continuous rasters. PCRaster has a few more data types. The following are all the data types used for PCRaster layers:

- Boolean: cells are 0 (False) or 1 (True).
- Nominal: cells have positive integer values representing classes (discrete raster) without a specific order, for example land-use classes.
- Ordinal: cells have positive integer values representing classes (discrete raster) with a specific order, for example stream order.
- Scalar: cells have real values (decimal, positive, negative). This is used for continuous rasters.
- Directional: for rasters with a compass direction in degrees or radians. For example, aspect.
- LDD: local drain direction, a specific data type for flow direction rasters.

During calculations, PCRaster checks the data type of the input rasters. Calculations can only be done when the correct data types are used.

Here we will convert the dem_subset layer to the PCRaster format.

16. In the *Processing Toolbox* go to PCRaster | Data management | Convert to PCRaster Format.

17. In the *Convert to PCRaster Format* dialog, choose dem_subset as *Raster layer* and Scalar as *Output data type* (the DEM is a continuous raster, therefore we use the scalar data type here). Save the result to dem.map (figure 4.23).

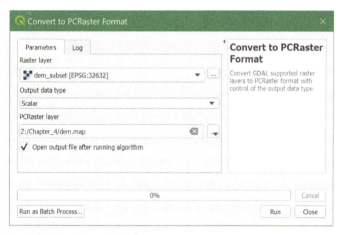

Figure 4.23: Convert to PCRaster Format Dialog

18. Click *Run*.

The result is now added to the map canvas

19. Click *Close* to close the dialog.

Calculate the Flow Direction

The next step is to fill the sinks (artificial depressions in the DEM) and calculate the flow direction raster. PCRaster does both in one step with the *lddcreate* tool.

20. In the *Processing Toolbox* go to PCRaster | Hydrological and material transport operations | lddcreate.

21. In the *Lddcreate* dialog, choose dem as *DEM layer*. Keep all other defaults so it will fill the sinks completely.

With the *lddcreate* tool you can control the filling by changing these parameters. Click the documentation link in the tool for more information. Save the results to `flowdirection.map` (figure 4.24).

Figure 4.24: The Lddcreate Dialog

22. Click *Run*. This can take a few minutes before the result appears in the map canvas. Click *Close* to close the dialog.

> The `lddcreate` tool fills the sinks in the DEM and results in a local drain direction (flow direction) map. If you need the filled DEM, you can use the `lddcreatedem` tool.

Styling the Flow Direction Layer using a Circular Color Ramp

A flow direction layer indicates the direction of flow for each pixel. Direction can be expressed as compass direction, however, we cannot store text in a raster. Compass direction can also be expressed as degrees on a circle where north is 0 degrees, east is 90 degrees, etc. To store 360 degrees, we would need more than 8 bits (range from 0 to 255), which would increase the file size. In addition, the *D8* method uses discrete directions to the 8 surrounding cells. Therefore, GIS software recodes the 8 directions. Each software, however, does it in their own way. The PCRaster LDD format uses the values of the numeric keypad for the directions.

Besides knowing the encoding of the compass directions in the raster, we also need to apply a directional or circular color ramp instead of a linear one that we have used until now for continuous rasters. In this section, we will define a circular color ramp to intuitively show southern oriented flows with warm colors and northern facing flows with cool colors.

We will use the *Raster Attribute Table* for that.

23. Open the *Layer Styling* panel for the flowdirection layer.

24. Change the renderer to Paletted/Unique values, because the flow directions are encoded in discrete values from 1 to 9. Click *Classify* to assign random colors to the cell values.

25. Right-click on flowdirection in the *Layers* panel and choose New Attribute Table.

26. In the *New Attribute Table* pop-up choose the default *GDAL auxiliary XML* format and click *OK*.

27. Click *Open Raster Attribute Table* in the message at the top of the map canvas.

28. Toggle on the editing of the attribute table and edit the values in the *Class* field to reflect the compass directions as text.

29. Change the colors in the *RAT Color* field by clicking on the colors and using the RGB values from figure (4.25).

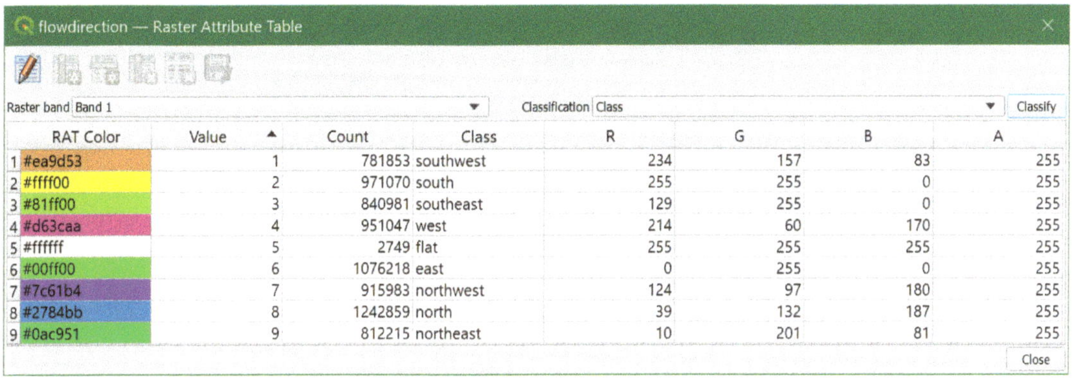

RAT Color	Value	Count	Class	R	G	B	A
1 #ea9d53	1	781853	southwest	234	157	83	255
2 #ffff00	2	971070	south	255	255	0	255
3 #81ff00	3	840981	southeast	129	255	0	255
4 #d63caa	4	951047	west	214	60	170	255
5 #ffffff	5	2749	flat	255	255	255	255
6 #00ff00	6	1076218	east	0	255	0	255
7 #7c61b4	7	915983	northwest	124	97	180	255
8 #2784bb	8	1242859	north	39	132	187	255
9 #0ac951	9	812215	northeast	10	201	81	255

Raster band Band 1   Classification Class   Classify

Figure 4.25: Flowdirection Raster Attribute Table

30. Toggle off the editing and save the changes. Click *Classify* to apply the changes and click *Close* to close the attribute table.

When you blend the flowdirection layer with the hillshade layer the result looks like figure 4.26, on the next page.

Styling the Flow Direction Layer using Arrows

We can further improve the styling of the flowdirection layer by adding arrows. This can be done using the *mesh* styling functionality of QGIS. To use that functionality, we need to convert the PCRaster LDD to a mesh format. We can do that with the *Crayfish* plugin.

31. Install the *Crayfish* plugin from the *Plugins Manager* (figure 4.27, on the following page).

32. In the *Processing Toolbox*, go to Crayfish | Conversions | PCRaster LDD to GRIB.

33. In the *PCRaster LDD to GRIB* dialog, choose flowdirection as *Input raster* and flowdirmesh.grb as *Output file (GRIB)* (figure 4.28, on page 95).

34. Click *Run*.

35. In the *Browser* panel, expand the flowdirmesh.grb group (you might need to refresh the

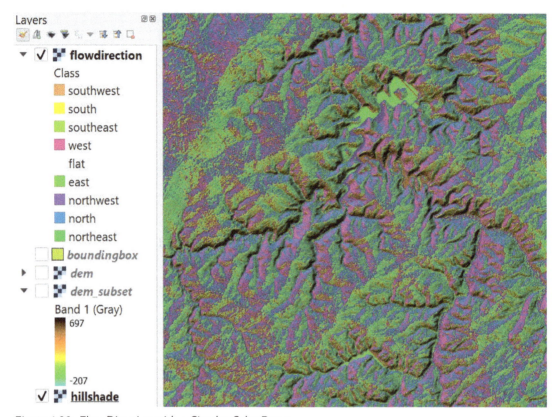

Figure 4.26: Flow Direction with a Circular Color Ramp

Figure 4.27: Installing the Crayfish Plugin

Figure 4.28: PCRaster LDD to GRIB Dialog

Browser panel with the 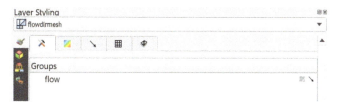 button) and drag the flowdirmesh layer with the mesh icon to the map canvas.

This might take some time. If the file is too large for your computer's memory, you can get errors. In that case, you can clip the flowdirection layer to a smaller area and repeat the steps to convert the file to the mesh format.

When the map canvas shows a completely yellow layer, the flowdirmesh layer has been loaded and we can start styling it.

36. Select the flowdirmesh layer in the *Layers* panel and open the *Layer Styling* panel.

37. In the *Layer Styling* panel, go to the *Datasets* tab and click on to disable contours and click on the arrow to enable vectors (figure 4.29).

Figure 4.29: Mesh Datasets Tab

Now so many arrows are drawn in the map canvas that it turns black. Let's tune the settings to improve this.

38. Go to the *Vectors* tab (figure 4.30, on the next page) and change the *Arrow Length* settings to Fixed and the *Length* to 2.00. Change the *Color* to dark blue.

39. Zoom in to see the flow direction with arrows.

40. Check the box to *Display on User Grid* to show the arrows fixed to a grid, e.g. with an *X and Y Spacing* of 10 px (figure 4.30, on the following page).

The settings are depending on your zoom level. Play with the settings to get a nice result. You can also try the other *Symbology* settings for visualization as Streamlines and Traces.

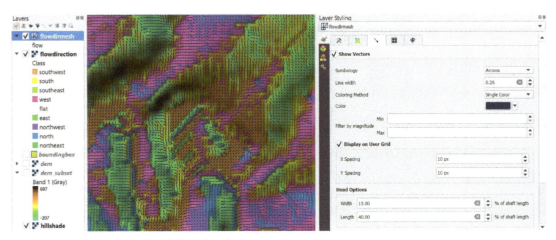

Figure 4.30: Flowdirection Mesh Styled with Arrows

Visualize Flow Direction in 3D

We can also visualize the flow direction using the QGIS *3D Map View*.

41. In the main menu, choose View | New 3D Map View (figure 4.31).

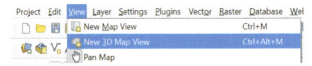

Figure 4.31: New 3D Map View in the Menu

42. In the *3D Map* view, click the *Options* button and choose Configure....

43. In the *3D Configuration* dialog, stay in the *Terrain* tab and change the *Type* to DEM (Raster Layer) and choose dem as *Elevation*. Click *OK* to apply the settings and return to the *3D Map* view.

The 3D view will now start rendering. Try to navigate the scene with the different mouse buttons and the compass. You can change the *Vertical scale*, *Tile resolution* and *Skirt height* in the *3D Configuration* to improve the visualization (figure 4.32, on the facing page).

4.7 Delineate Streams

Calculate Strahler Orders

Before we can derive the streams from the DEM, we need to determine what we consider streams. For this purpose, we use the Strahler order. The higher the order, the bigger the stream (see figure 4.33, on the next page).

1. Close the *3D Map* view and show only dem_subset with blended hillshade by unchecking the other layers in the *Layers* panel. Also zoom out to the full extent.

2. Search for streamorder in the *Processing Toolbox* and select PCRaster | Hydrological and transport operations | streamorder (figure 4.34, on page 98).

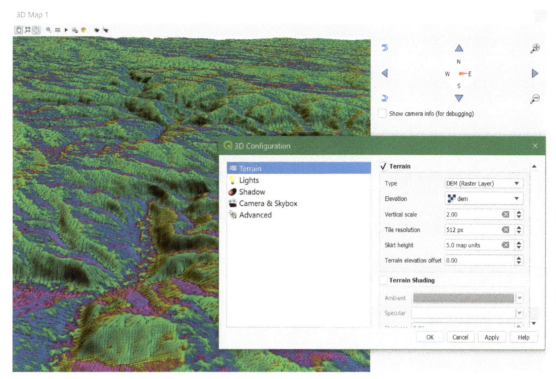

Figure 4.32: Flow Direction in the 3D View

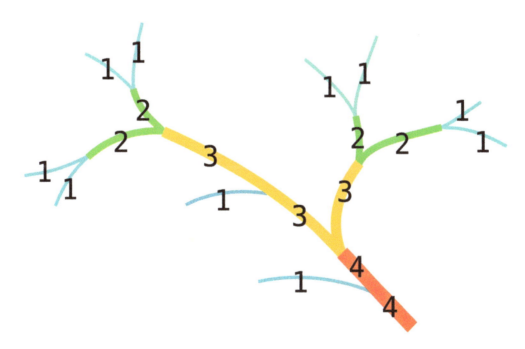

Figure 4.33: Strahler Order Method (Image from
https://commons.wikimedia.org/wiki/File:Flussordnung_(Strahler).svg, License CC BY-SA 3.0)

Figure 4.34: PCRaster Streamorder Tool

3. In the *Streamorder* dialog, select flowdirection for the *Local Drain Direction layer*, use strahler.map as the *Stream Order* output filename, and click *Run*. Click *Close* when the algorithm is done (figure 4.35).

Figure 4.35: Streamorder Dialog

To make more sense of the strahler layer, we are going to style it.

- Is the Strahler order layer a boolean, discrete, or continuous raster?

The Strahler order layer is a discrete raster, but the order of the classes is important. Therefore, for PCRaster, it has the ordinal data type. For discrete rasters in QGIS, we use the *Paletted/Unique values* styling. The higher the Strahler order, the bigger the stream. So we will use a color ramp from white to blue.

4. Open the *Layer Styling* panel if you closed it before and make sure that the strahler layer is selected.

5. Choose *Paletted/Unique values* from the drop-down menu.

6. Click *Classify* (figure 4.36, on the facing page).

This assigns a unique, random color to each unique value in the raster.

7. Right-click on *Random colors* and choose *Blues* (figure 4.37, on the next page).

8. Zoom in on the map canvas to evaluate the result.

The strahler layer now shows in an intuitive way that the higher orders will be larger streams than the lower orders (figure 4.38, on page 100).

Figure 4.36: Styling of Strahler Layer

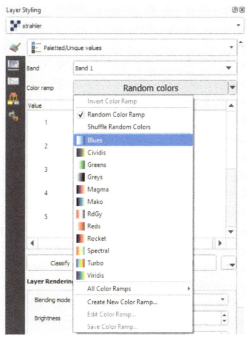

Figure 4.37: Select Blues Color Ramp

Calibrate Strahler Threshold to Determine Streams

The next step is to apply a calibration procedure to determine which Strahler orders we consider to be streams. We will create boolean layers for Strahler orders larger than or equal to a threshold value. Each boolean layer will be compared with a reference layer. Here we will compare the Strahler orders with the rivers on OpenStreetMap. For areas where there is a lack of data in OpenStreetMap, you could use the Google Satellite. Both OpenStreetMap and Google Satellite layers can be found in the *QuickMapServices* plugin.

9. Add an OpenStreetMap backdrop layer from the QuickMapServices by choosing `Web | QuickMapServices | OSM | OSM Standard` from the main menu. Make sure that only the `OSM Standard` and `strahler` layers are visible.

Figure 4.38: Strahler Orders Styled with Intuitive Colors

We will use the *Raster Calculator* to create a boolean map with 1 (True) for Strahler order >= 5 and 0 (False) for the other values.

10. Go to `Raster | Raster Calculator`.

11. Fill in the dialog as in figure 4.39, on the next page. Note that we check the box to *Create on-the-fly raster instead of writing layer to disk*, because we only need these layers for calibration purposes. Choose `strahler >= 5` as the *Layer name* so we can compare results later. Click *OK* to run the calculation.

It is also good practice to style the `strahler >= 5` layer.

12. Hide the `strahler` layer in the *Layers* panel so we have only the `strahler >= 5` and `OSM Standard` layers checked.

13. Go to the *Layer Styling* panel and make sure that `strahler >= 5` is selected.

14. Choose *Paletted/Unique values* from the drop-down menu.

15. Click *Classify*.

This layer is boolean and therefore it has only pixels with values 0 and 1, but it also shows a nodata value. For our calibration it is important to make the ones blue and the rest of the pixels transparent so we can compare the raster with the streams on OpenStreetMap.

16. Click on the color for value 1 and change it to dark blue. Select the two other classes with the `Ctrl` button pushed. Then click the *Remove selected row(s)* button to remove these classes (figure 4.40, on the facing page).

On the map canvas, we can now see all streams larger than or equal to Strahler order 5 and we can compare them with the rivers on the OpenStreetMap.

17. Repeat the steps in this subsection for different Strahler order threshold values and determine the one that corresponds best with the rivers on OpenStreetMap. Remove the other boolean layers.

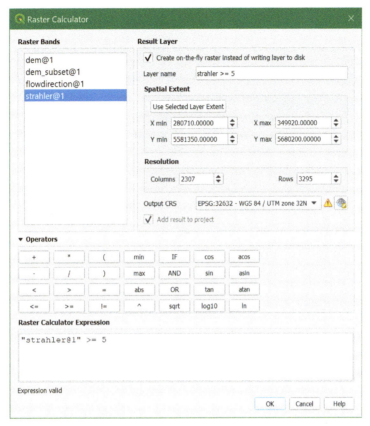

Figure 4.39: Calculate a Boolean Map with the Raster Calculator

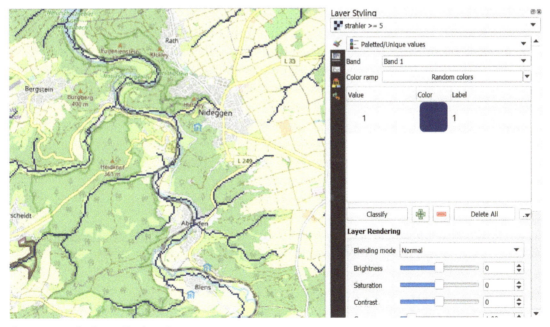

Figure 4.40: Styling a Boolean Layer

Tip: You can copy the styles of the layers: right-click on a layer, choose Styles | Copy Style and then right-click on the target layer and choose Styles | Paste Style (figure 4.41).

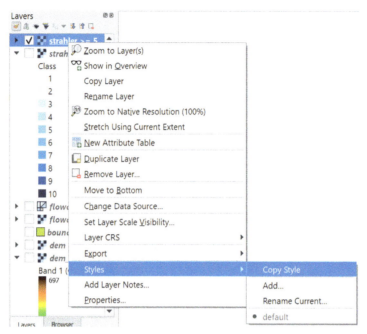

Figure 4.41: Copy the Style of a Layer

Calculate the Channel Network

The next step is to calculate the channel network.

First, we are going to create a layer with the Strahler orders for the rivers by selecting the cells with order equal to or higher than the threshold determined in the previous step. We will use order 8 as the threshold.

With the *PCRaster Tools* plugin, we can only use raster layers in map algebra. If we want to calculate a new boolean layer with strahler >= 8 as True (our channels) and < 8 as False, we need to first create an ordinal raster that consists of cells with only value 8. This can then be used in map algebra.

18. In the *Processing Toolbox*, go to PCRaster | Data management | spatial.

19. In the *Spatial* dialog, enter 8 for *Input nonspatial*. Choose Ordinal for *Output data type* and strahler as the *Mask layer*. Call the *Output raster layer* ordinal8.map (figure 4.42, on the facing page).

20. Click *Run* and *Close* the dialog after completion.

Now we are going to create a boolean layer with 1 (True) for all strahler cells that are >= 8.

21. In the *Processing Toolbox*, go to PCRaster | Conditional and boolean operators | comparison operators.

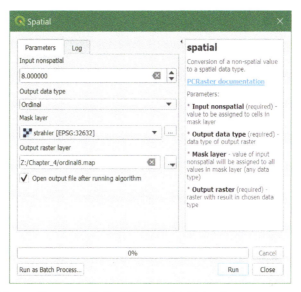

Figure 4.42: Spatial Dialog to Create an Ordinal Raster with Value 8

22. In the *Comparison Operators* dialog, choose `strahler` as *Input raster,* `>=` as *Comparison operator* and `ordinal8` as the second *Input raster.* Save the *Output Boolean raster* as `channels.map` (figure 4.43).

Figure 4.43: Comparison Operators Dialog

23. Click *Run* and *Close* the dialog after completion.

Now we have a boolean layer with the channels that we can use to assign the river Strahler orders.

24. In the *Processing Toolbox,* go to `PCRaster | Conditional and boolean operators | ifthen`.

25. In the *If Then* dialog, choose `channels` as the *Input Boolean Condition Raster* and `strahler` as the *Input True Raster.* Call the *Output Raster* `channelsstrahler.map` (figure 4.44, on the following page).

Figure 4.44: If Then Dialog

This means: if the channels layer has cells that are True (1), then give those cells the value of the strahler layer. All other cells get "nodata".

26. Click *Run* and *Close* the dialog after processing.

Now we have a raster with the Strahler orders for the river channels.

Convert Raster to Vector Lines

For a better visualization of the channel network, we need to convert the channelsstrahler layer to a line vector.

To do this, we will use the *GRASS* tools from the *Processing Toolbox*.

First, we need to thin the raster lines so they are only one cell wide.

27. From the *Processing Toolbox*, choose GRASS | Raster (r.*) | r.thin.

28. In the *r.thin* dialog, choose channelsstrahler as *Input raster layer to thin*. Keep the defaults and save the *Thinned* raster as channelsthin.tif. Make sure to choose the GeoTIFF format instead of the PCRaster .map format (figure 4.45, on the next page).

29. Click *Run* to perform the thinning and click *Close* when the processing is completed.

30. Compare the result with the channelsstrahler layer to understand what *r.thin* does.

Now we can convert the channelsthin layer to a line vector.

31. In the *Processing Toolbox*, go to GRASS | Raster (r.*) | r.to.vect.

32. In the *r.to.vect* dialog (figure 4.46, on page 106):

   • choose channelsthin as *Input raster layer*.
   • choose line from the drop-down menu as *Feature type*.

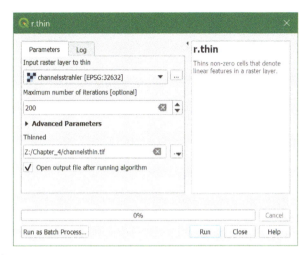

Figure 4.45: r.thin Dialog

- check the box to *Use raster values as categories instead of a unique sequence.*
- under *Advanced Parameters*, change the *v.out.ogr output type* to `line`.
- name the *Vectorized* output `channels.shp`.

33. Keep the rest as default. Click *Run* to perform the conversion. Click *Close* when the processing is done.

> The result has some geometrical errors that can be fixed using v.clean from GRASS that you can find in the Processing Toolbox or with the Topology Checker plugin. This is beyond the scope of this book.

Styling the River Channels

Now we can style the `channels` vector layer to get a better understanding of the results. We will apply a styling in such a way that the lines get thicker with higher Strahler orders.

34. Open the attribute table of the `channels` layer.

In the attribute table, you can see that the *cat* field has the original Strahler order values (in our case 8, 9 and 10). In reality, the Strahler order method starts with value 1 for the smallest tributaries. Therefore, we need to reclassify the values in the *cat* field to represent the real Strahler orders.

35. Click the *Open field calculator* button.

36. In the *Field Calculator* dialog, create a new field with the name `ORDER`. It should have an *Output field type* of `Whole number (integer)` and an *Output field length* of 1. Under *Expression*, write the code as given in figure 4.47, on page 107).

The `CASE` expression is used to evaluate a series of conditions. In our case, that means that it first checks if `cat` is equal to 8. If that's the case, it assigns value 1, etc. The `CASE` expression ends with the `END` statement.

37. Click *OK* to apply the calculation.

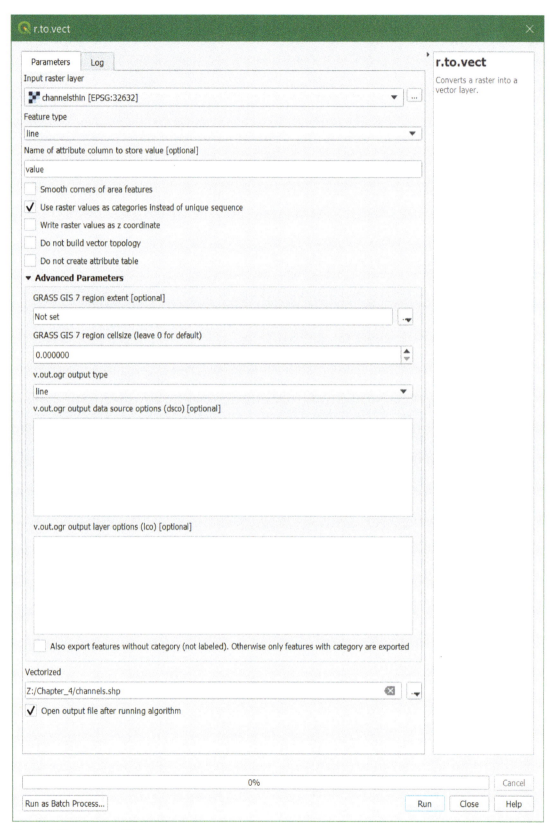

Figure 4.46: r.to.vect Dialog

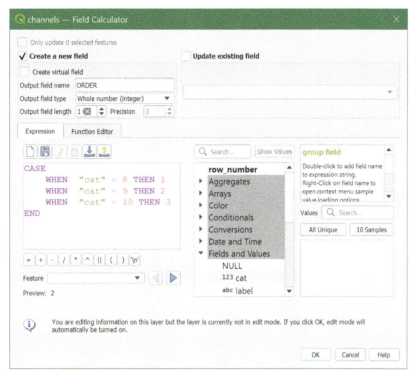

Figure 4.47: Calculate Strahler Field with the Field Calculator

38. Check the result in the attribute table. Toggle off editing by clicking ✏ and confirm that you want to save the edits. You can now close the attribute table.

39. Open the *Layer Styling* panel and set the target layer to *channels*.

40. Keep the *Single Symbol* renderer, but change the *Symbol layer type* to `Interpolated Line`.

41. Change the *Stroke Width* to `Varying Width`. Set the *Start value* and *End value* to the `ORDER` field. Click the 🔄 button to get the correct *Min. value* and *Max. value*. Change the *Min. width* to `0.30` Points and the *Max. width* to `1.00` Points (figure 4.48, on the following page).

42. Under *Color*, keep the `Single Color` but change it to blue with an RGB value of 15 | 66 | 220.

43. Your map should now resemble that in figure 4.49, on the next page, showing the streams identified as having higher Strahler Orders are the main channels and the smaller ordered streams are tributaries.

4.8 Define Outflow Point

A catchment is an extent or an area of land where surface water from rain, melting snow, or ice converges to a single point at a lower elevation, usually the exit of the basin, where the waters join another water body, such as a river, lake, reservoir, estuary, wetland, sea, or ocean. In order to delineate a catchment we need to have:

- the coordinates of our outlet in the same coordinate system as the layers we are using
- the channel network that matches the flow directions as calculated from a hydrologically

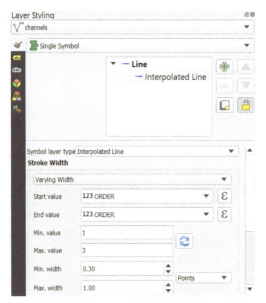

Figure 4.48: Styling Channel Strahler Orders using the Interpolated Line Renderer

Figure 4.49: Channels Data Styled by Strahler Order

correct DEM

The outflow point of the Rur catchment is in Roermond, where the Rur enters the Meuse river (Maas in Dutch). The channel network that has been derived in the previous step is in the channels layer. We will however use the channelsstrahler raster layer, because we need to define the outlet exactly on a river pixel.

1. Make sure you have the channelsstrahler layer on top of the OSM Standard layer from the *QuickMapServices* plugin. You can style the channelsstrahler layer with a blue ramp, so that the main channel appears in dark blue.

2. Look for the location where the Rur river flows into the Meuse. Note that on OpenStreetMap the Dutch names are used, because it lies in the Netherlands. Rur is spelled as Roer and Meuse is spelled as Maas.

> Note that in some places, the delineated channels are not corresponding well with the channels on OpenStreetMap. This can be for the following reasons: (1) Incorrect automatic delineation of streams, which can be caused by errors in the DEM or areas that are too flat, (2) Distortion due to reprojection and resampling (3) Human influence on the natural course of the channels. The catchment delineation, however, only works when the outlet is defined on a delineated channel, because that corresponds with our flow direction layer. Results can be improved by using different (parameters of) fill sinks algorithms or burning an existing vector layer of the stream network in the DEM.

3. Choose a pixel on the delineated channel that is close to the real outlet of the Rur in the Meuse (step 2).

4. Right-click on the pixel and copy the *Map CRS* coordinate (figure 4.50).

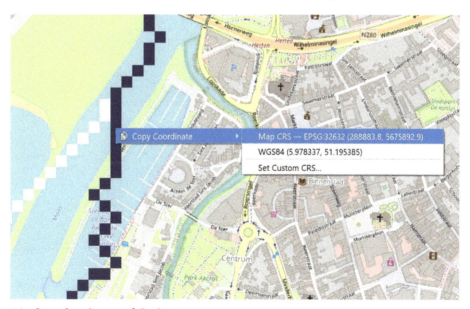

Figure 4.50: Copy Coordinates of Outlet

5. Paste the coordinates in a text editor (for example Notepad) and add a comma and a 1 (figure 4.51, on the next page).

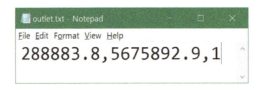

Figure 4.51: Outlet CSV File

> In our case, we want to delineate only the entire Rur catchment and give all pixels belonging to that catchment ID = 1. If you want to delineate multiple (sub)catchments, you can add more coordinates to the CSV file. Just make sure that you also add unique ID numbers.

6. Save the file as outlet.txt.

We can now convert the coordinate to a pixel in PCRaster format.

7. In the *Processing Toolbox,* go to PCRaster | Data management | Column file to PCRaster Map.

8. In the *Column File to PCRaster Map* dialog, browse to the text file with the outlet, choose flowdirection as the *Raster mask layer,* choose Nominal (small) as the *Output data type,* and save the *PCRaster layer* as outlet.map (figure 4.52).

Figure 4.52: Column File to PCRaster Map Dialog

9. Click *Run* and *Close* after processing.

Now the outlet point has been added to the map canvas.

10. Style the pixel with a clear color using the *Paletted/Unique values* renderer (figure 4.53, on the next page).

4.9   Delineate the Catchment

Now we're going to use the defined outlet to calculate the upstream contributing area (i.e. catchment) that produces discharge at this location.

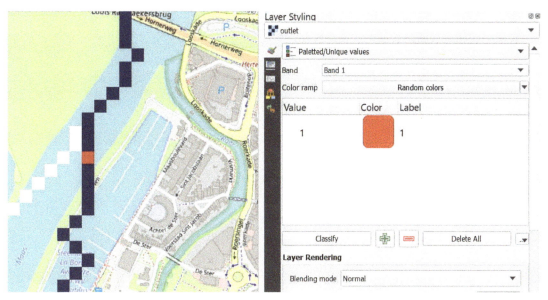

Figure 4.53: Styled Outlet

1. In the *Processing Toolbox,* go to `PCRaster | Hydrological and material transport operations | catchment`.

2. In the *catchment* dialog, use `flowdirection` as the *LDD layer,* use `outlet` as the *Outlet layer,* and save the result to `catchment.map` (figure 4.54).

Figure 4.54: Catchment Dialog

3. Click *Run* and *Close* the dialog after processing.

The result should look like figure (4.55, on the next page) when you zoom to the extent of the `catchment` layer. The result is a nominal raster with a cell value 1 for the catchment and 0 for outside the catchment. Nominal, because our outlet raster was defined as nominal data type. If the `outlet.txt` file had more coordinates of outlets, each catchment would have had cells with the value of the corresponding outlet cell. If you have subcatchments with outlets upstream of a higher order catchment, you can use the PCRaster *subcatchment* tool to avoid the overlap.

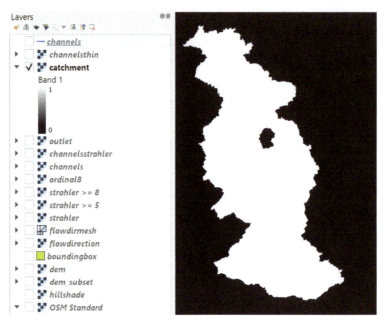

Figure 4.55: Catchment Raster Result

In order to overlay the catchment boundary with other data, it is better to convert it from raster to vector (polygon).

4. To convert the raster layer to vector, go to the main menu and choose `Raster | Conversion | Polygonize (Raster to vector)` (figure 4.56).

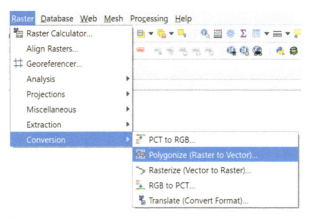

Figure 4.56: Polygonize Menu

5. Make sure you choose the right input and call the output `Rur_catchment.shp`. Click *Run*. Click *Close* to get back to the main screen (figure 4.57, on the next page).

6. Look at the result. Also check the attribute table (right-click on layer name and choose `Open attribute table`).

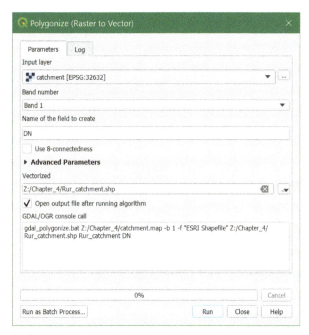

Figure 4.57: Polygonize Dialog

In the *catchment* calculation, cells belonging to the catchment get a value of 1, while the other cells get a value of 0. During the conversion to polygons, it can happen that geometry errors are introduced. If you find more than one feature with a value of 1 this indicates a geometry error (incorrect topology), because the boundary of the polygon makes a loop (see figure 4.58). This can give errors when we use the polygon for geoprocessing. You can fix those errors with the *fix geometry* tool in the Processing Toolbox.

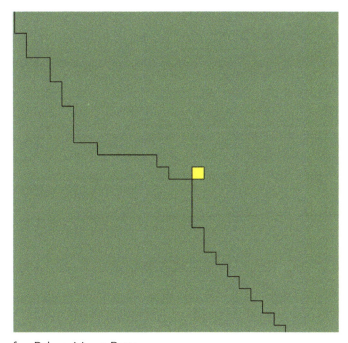

Figure 4.58: Issues after Polygonizing a Raster

Of course, we are only interested in the catchment area, so we have to remove the outside polygon.

7. In the attribute table, toggle to editing mode using the ✏ button, then select the record that you want to remove by clicking on the row. The selection will be highlighted in yellow on the map. Click the 🗑 button to delete the selected feature (figure 4.59).

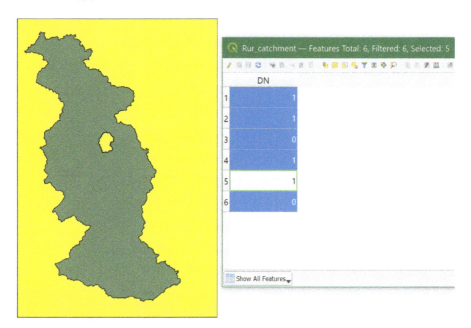

Figure 4.59: Delete Polygons that are not the Rur Catchment

You might have noticed that the polygon has a hole. This is caused by a huge open pit lignite mine, which results in its own subcatchment in our procedure. However, we would like to see the mine as part of the delineated catchment and will correct this.

8. Enable the *Advanced Digitizing Toolbar* that we have also used in Chapter 1 (right-click on a toolbar and check the box before the toolbar).

9. Click the *Delete Ring* 🔗 button and click on the mine.

The hole has now disappeared.

10. Toggle off editing by clicking ✏ again, and save the changes.

11. Now remove all unnecessary layers from the layers list so that we have only channels, Rur_catchment, dem_subset, hillshade and OSM Standard (in that order).

4.10   Clipping Layers to the Catchment Boundary

For visualization, it is nicer to clip the layers to the boundary of the catchment.

12. Let's first clip the channels vector layer to see only the streams that are inside the catchment. Go to the main menu and select Vector | Geoprocessing Tools | Clip (figure 4.60, on the facing page).

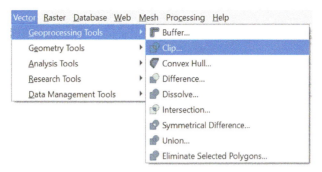

Figure 4.60: Clip Vectors Menu

13. Fill in the dialog as in figure 4.61 to use the catchment layer as a "cookie cutter" to clip the *Input layer* channels to the boundary of the *Overlay layer* Rur_catchment. Call the *Clipped* layer Rur_channels.shp. Click *Run* to run the tool. Click *Close* to return the main screen.

Figure 4.61: Clip Vectors Dialog

We can easily copy the styles from channels to Rur_channels.

14. Right-click on channels and choose Styles | Copy Style | All Style Categories.

15. Now right-click on Rur_channels and choose Styles | Paste Style | All Style Categories.

16. Remove channels from the layers list and make sure that Rur_channels layer is at the top of the list.

We can clip the DEM in the same way as we did in Create a Subset of the DEM, on page 82. The only difference is that we have to give a no data value that is out of the range of elevations—we often use -9999 for that. Here we don't need the clipped DEM, because we are going to apply a styling technique to create a nice effect to highlight our study area.

4.11 Styling the Resulting Catchment Area

To show the results of your analysis, you can use a technique named *Inverted Polygon Shapeburst Fills* to focus attention on the Rur Catchment.

1. Open the *Layer Styling* panel by clicking the ✎ button. Set the target layer to *Rur_ Catchment*.

2. Change from a *Single symbol* renderer to an *Inverted polygons* renderer. This renders the data as the inverse of its geometry. This creates a mask around the Rur valley.

3. Next, select the *Simple Fill* component. Change the *Symbol layer type* to *Shapeburst fill*. In the *Gradient colors* section, use the default *Two color* method. Change the first color to an RGB value of 135 | 135 | 135. Change the second color to white with an opacity of 65%.

4. In the *Shading style* section, click the *Set distance* option and set the distance to 4 mm and increase the *Blur strength* to 10.

5. Finally, at the top of the *Layer Styling* panel, select the *Shapeburst fill* component.

6. Click the *Add symbol layer* ⊕ button. Change the new *Simple fill* renderer to a *Symbol layer type* of *Outline simple line*. Give it a *Color* of black and a *Stroke width* of 0.46 mm.

This gives us the nicely styled basin as shown in figure 4.62.

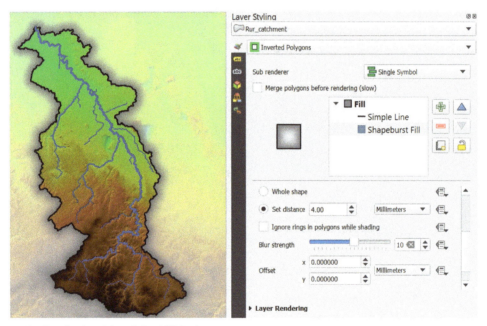

Figure 4.62: Rur Basin with a Color Hillshade

4.12 Storing the Data in a GeoPackage

To keep the data together and enable easy distribution, it is good to save the layers as a GeoPackage.

1. In the *Processing Toolbox* look for the *Package Layers* tool.

2. In the *Package Layers* dialog, click the `...` and select all layers. Note that these are only the vector layers (figure 4.63, on the facing page).

3. Save it as Rur_data.gpkg and click *Run* and *Close* when it's done. By default, it will also save

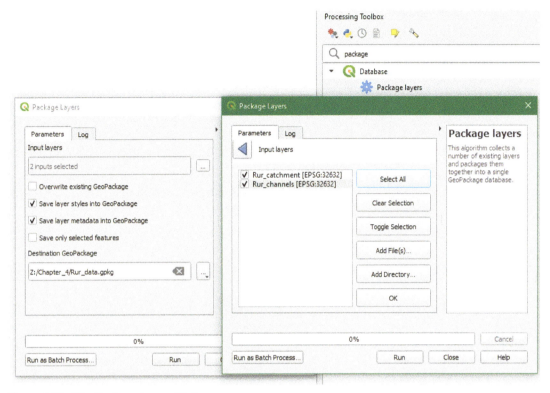

Figure 4.63: The Package Layers Dialog

the styles.

4. We can add the raster layers from the *Browser* panel. Simply drag the raster layers (in our case dem_subset) to the Rur_data.gpkg. You might need to refresh the *Browser* panel to see the GeoPackage.

Finally, we can also save the project in the GeoPackage. In that way, all data and settings are stored in one file that can be shared with others in a much easier way than separate shapefiles, GeoTIFFs, and styling files.

5. Add the layers from the GeoPackage to the project by dragging them from the *Browser* panel to the map canvas.

6. Copy the styles from dem_subset and hillshade to their respective layers from the GeoPackage. If you get confused, you can hover your mouse over the layer names to see the paths where they are stored.

7. Remove the layers that are not stored in the GeoPackage and rename the remaining layers to their original names.

Now we can save the project to the Rur_data.gpkg with the correct paths.

8. In the main menu, choose Project | Save to | GeoPackage... (figure 4.64, on the next page).

9. In the pop-up, browse to the Rur_data.gpkg at *Connection* and name the *Project* Rur. Click *OK* to save the project in the GeoPackage.

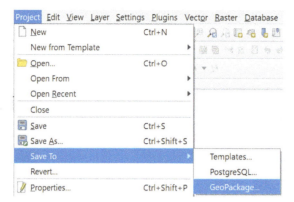

Figure 4.64: Save Project to GeoPackage Menu

The end result should now look like figure 4.65.

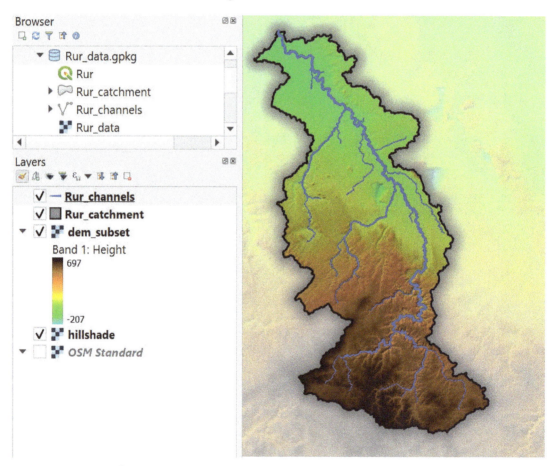

Figure 4.65: Result of Chapter 4

5. Adding Open Data to Your Catchment

5.1 Introduction

Now that we have defined the boundaries of our study area, the Rur catchment, we can look for open data that is available on the Internet.

After this chapter you will be able to:

- add *OGC web map services* to QGIS
- download vectors from *OpenStreetMap* using the *QuickOSM plugin*
- make *temporary scratch layers* permanent
- style open data from the web

In this exercise we will use:

- ESA WorldCover data https://esa-worldcover.org/en/data-access
- Data from OpenStreetMap: http://www.openstreetmap.org

Instead of using a web browser, we will create a live link between the online data and our data in QGIS, using an OGC web map service, an open standard for sharing maps through the Internet. We will also download vector layers from OpenStreetMap through the QuickOSM plugin in QGIS.

For this exercise, you need the delineation of the Rur catchment and its streams from the previous chapter.

Before we start, make sure that you have at least the following layers loaded into QGIS:

- Boundary of the Rur catchment styled with the inverted polygon shapeburst fill
- Streams in the Rur catchment
- OSM Standard backdrop from the QuickMapServices

There are several ways to begin this new map document while maintaining the symbology established during the catchment delineation exercise. You can open the map document from the previous chapter and choose Save As. You can also open a new map document and copy/paste the layers from the catchment delineation exercise into it. To do this, have the project open from the previous chapter, select the layers, right-click, and choose Copy Layer from the context menu. Then right-click in the Layers Panel for the new project and choose Paste Layer/Group.

You can hide the DEM and hillshade layer from the previous project. Your map should now look like figure 5.1, on the next page.

For this chapter, the videos from the *QGISHydro Chapter 5* playlist at YouTube channel http://www.youtube.com/c/hansvanderkwast provide the theoretical background, results of the steps, and additional materials.

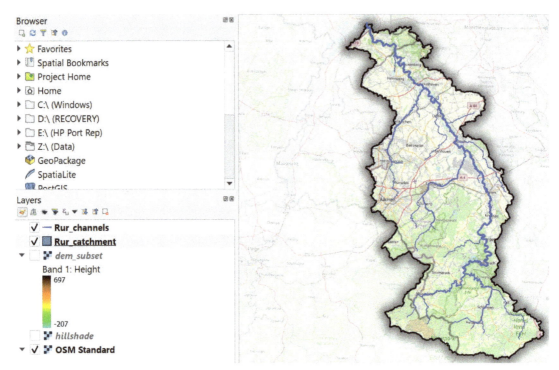

Figure 5.1: The Map at the Start of Chapter 5

5.2 Adding Data from OGC Web Map Services

In this section, we are going to use open data from a web map service—the ESA World-Cover product (Zanaga et al., 2021 - `https://doi.org/10.5281/zenodo.5571936`). The ESA WorldCover is a global land cover map for 2020 at 10 m resolution. It has been derived from Sentinel-1 and Sentinel-2 data. The different ways to access the data are described here: `https://esa-worldcover.org/en/data-access`.

1. Go to the website `https://esa-worldcover.org/en/data-access` and scroll down to the *Web Map Services* section.

Here you can see that you can add the land cover map as WMTS or WMS layer. Both deliver a rendered picture of the data and not the data itself. Therefore, it can be used as a backdrop, but not for analysis. A WMTS delivers the data in tiles, while the WMS delivers the whole map. We'll use the WMTS web map service.

2. Go to your QGIS project and click the *Open Data Source Manager* button ![icon] from the toolbar under the main menu.

3. In the *Data Source Manager* dialog, choose WMS/WMTS.

4. In the dialog that opens, click the *New* button.

5. In the dialog that follows, type `ESA WorldCover 2020` for *Name*.

6. On the WorldCover data website, right-click on the WMTS link and choose `Copy Link` (figure 5.2, on the facing page).

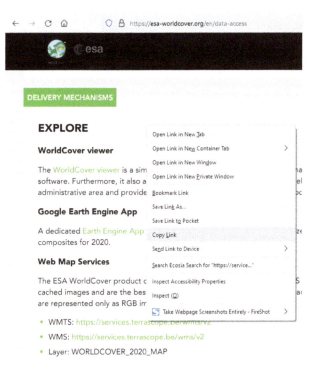

Figure 5.2: Copy Link from Web Page

7. Paste the link in the *Create a New WMS/WMTS Connection* dialog in QGIS (figure 5.3, on the next page). Keep the defaults and click *OK*.

8. Back in the other dialog window, click *Connect*. The layers will now be retrieved from the WMTS server (figure 5.4, on page 123).

There are many layers available from this web map service that you can now see under the *Tilesets* tab.

9. Click on WORLDCOVER_2020_MAP so that it is highlighted and click *Add* and *Close* to return to the main screen.

10. Drag the layer below the Rur_catchment layer so you can clearly see the area within the Rur catchment.

In this case, there is no legend provided with the web map service. You can manually add a legend.

11. Go back to the web page with the WorldCover data.

12. Click on *WorldCover viewer*.

This is an online viewer for the WorldCover data.

13. Make a screenshot of the legend and save it as worldcoverlegend.png.

14. In QGIS, right-click on the WORLDCOVER_2020_MAP layer and choose Properties.

15. In the *Layer Properties* dialog, go to the *Legend* tab.

Figure 5.3: Create a New WMTS Connection

16. At *Legend placeholder image,* browse to worldcoverlegend.png (figure 5.5, on the next page).

17. Click *OK* to apply and close the dialog.

Your window should now look like figure (5.6, on page 124).

Have a closer look at the land cover map.

- What are the largest differences in land cover between the upstream and the downstream areas of the Rur catchment?
- What are the largest cities in the catchment?
- Compare the land cover map with Google Satellite from the *QuickMapServices* plugin. Are there important land cover or land use features missing that influence the hydrology in the Rur catchment?

18. Save the project before you continue.

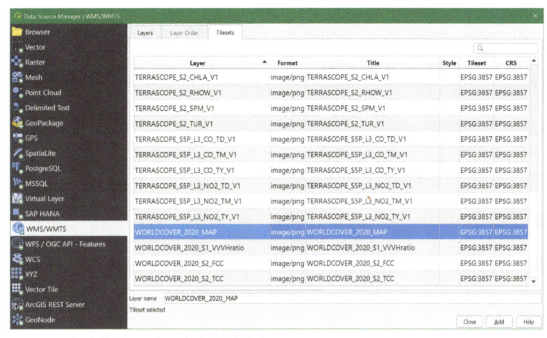

Figure 5.4: Available Layers through the WMTS Connection

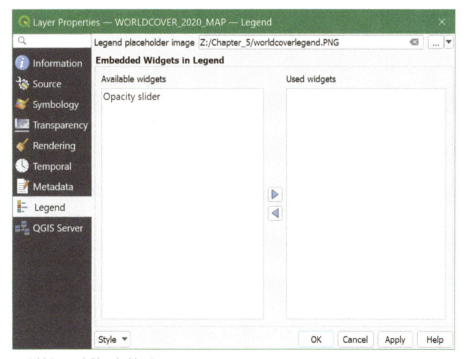

Figure 5.5: Add Legend Placeholder Image

Figure 5.6: WMTS Layer with Legend

There are different web services you can use in QGIS. There are the OGC services: WMS, WFS, and WCS. While WMS, WMTS render a picture from the data, WFS and WCS will give you the vector and raster data respectively. You can also connect to Spatial Data Infrastructures that use GeoNode or to ArcGIS Map Server and ArcGIS Feature Server. Also Vector Tiles can be added. With vector tiles you can control the styling and labeling of the image. With the MapTiler plugin you can add several useful vector tiles to QGIS. QGIS also supports XYZ Tiles. With the QuickMapServices plugin you have already used them. All these connections are available in the *Data Source Manager* and the *Browser* panel.

5.3   Adding Vector Data from OpenStreetMap

OpenStreetMap (OSM) is a collaborative project to create a free editable map of the world. OSM is considered a prominent example of volunteered geographic information (VGI) or crowd-sourcing. There are several ways to use the data:

- Through the interactive map on the OpenStreetMap website (`http://www.openstreetmap.org`).
- In QGIS, you can add OSM basemaps via the *QuickMapServices* plugin or drag it from the XYZ tiles in the *Browser* panel.
- In QGIS, you can download OSM data directly over the Internet. There are several ways to do this. In this section, we'll use the *QuickOSM* plugin which uses the Overpass API (`https://wiki.openstreetmap.org/wiki/Overpass_API`).

In this video, Etienne Trimaille, developer of the QuickOSM plugin, gives an overview of different ways of using OpenStreetMap data in QGIS: https://youtu.be/zCdbtoBctM4.

In this section, we are going to download OSM vector data directly in QGIS for the Rur Catchment. We continue from the previous results, but we only visualize the Rur_channels, Rur_catchment and OSM Standard layers. The other layers should be unchecked.

1. Install the *QuickOSM* plugin through the main menu: Plugins | Manage and Install Plugins.... Search for QuickOSM.

2. Open the *QuickOSM* dialog by choosing Vector | QuickOSM | QuickOSM... from the main menu (figure 5.7).

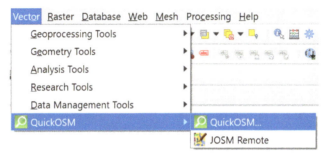

Figure 5.7: QuickOSM Menu

When you use *QuickOSM* from the first time, a pop-up will show up where you have to confirm that you understand the copyrights of OpenStreetMap, before using the plugin. In Chapter 7, Map Design, you will use the official way of crediting OpenStreetMap in your final map.

3. Click *I understand the copyrights, access to the plugin*.

We're first going to download the rivers so we can compare them with the rivers that we have previously derived. The OSM data attributes consist of keys and values. To learn more about it, check https://wiki.openstreetmap.org/wiki/Mapfeatures.

4. Choose *waterway* as *Key*, *river* as *Value*. From the drop-down menu, choose Layer Extent and Rur_catchment as the extent. Note that you can also select the extent of the Map Canvas. Click the arrow before *Advanced* and make sure only *Node, Way, Relation*, and *Lines* are checked (you always need to select the geometry that you would like to download, some features are available with multiple geometries). The dialog should now look like figure 5.8, on the following page. Click *Run Query*.

You can increase the *Timeout* value if you have a slow Internet connection.

The new layer will be added as a temporary scratch layer as indicated by the *Temporary Scratch Layer* icon in the indicator space to the right of the layer in the *Layers Panel*. This means that it will be discarded after closing QGIS. It is important to make these layers permanent if you want to keep them. We will do this later.

5. Close the *QuickOSM* dialog window.

6. Adjust the style and compare the waterway_river from OSM with the Rur_channels layer.

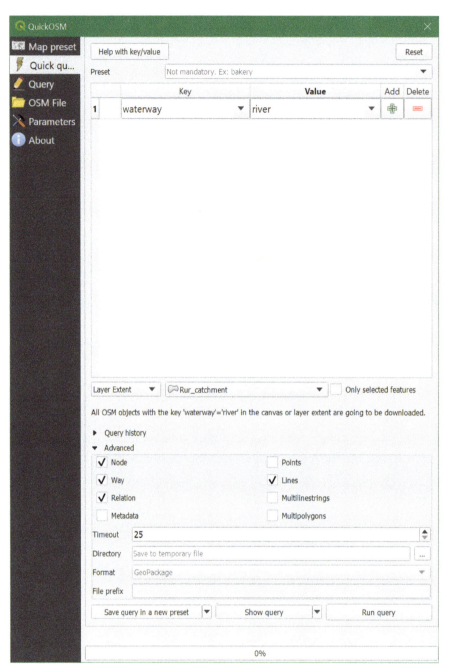

Figure 5.8: QuickOSM Dialog

- What do you observe?

7. Let's add the mines in a similar way. Use key=landuse and value=quarry. Don't forget to select *Multipolygons* instead of *Lines*.

8. Style the polygons with a gray fill and black stroke.

9. Label the polygons with the Name attribute and an italic font. Switch to the label *Formatting* tab and enter a space as the *Wrap on character*. Then set the *Alignment* to *Center*. Switch to the label *Rendering* tab and click *Only draw labels which fit completely within feature*.

Now we are going to look around a specific place. In the lower left of the QGIS window you can find in the status bar a search field, which is called the *Locator bar*. This can be used to find and run any feature or option in QGIS. You can also use it to lookup places using the Nominatim *geocoder*, which is made available by the OpenStreetMap Foundation and contributors. To lookup places type > followed by a space and the address. We are going to lookup the place Jülich.

10. In the *Locator bar* type > Jülich. It will show immediately the result from the geocoder (figure 5.9).

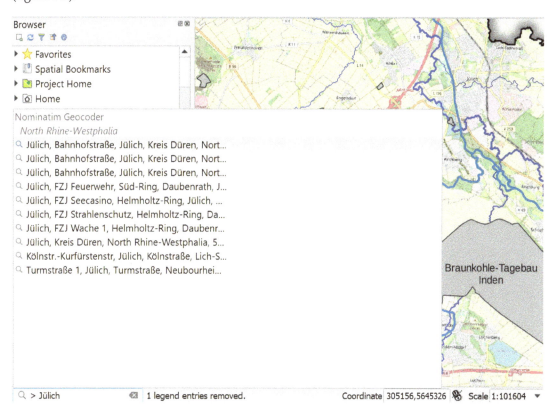

Figure 5.9: Geocoding from the Locator Bar

11. Choose *Jülich, Kreis Düren, North Rhine-Westphalia, 52428, Germany*.

QGIS now zooms in on the center of Jülich.

Southeast of Jülich is the Forschungszentrum Jülich, a large research institute. In the south and east we see a large surface mining of lignite. The one in the south (Braunkohle-Tagebau Inden) is in the Rur catchment.

12. Now compare the OSM derived quarry with a Google Satellite, the DEM, and the World-Cover land cover map.

- What are the differences?
- Which one is more up-to-date?
- How does the hydrography relate to the quarry (OSM versus GIS delineated)?

Remember that the layers that were added via *QuickOSM* are temporary scratch layers, indicated by 🗔 . There is an option to make the layer permanent, but the export layer option is more flexible. We will use that to save the layer to the GeoPackage created in the previous chapter (Storing the Data in a GeoPackage, on page 116).

13. Right-click on the `landuse_quarry` layer and choose `Export | Save Features As...` (figure 5.10).

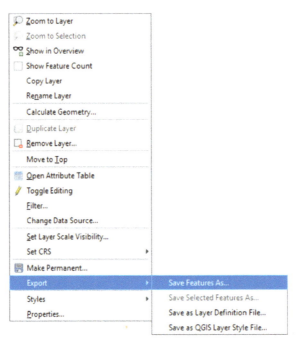

Figure 5.10: Export a Layer to Another Format and/or Projection

14. In the *Save vector layer as* dialog (figure 5.11, on the facing page), for *Format*, choose *GeoPackage*. At *Filename*, browse to `Rur_data.gpkg` created in the previous chapter. At *Layer name*, type `Quarries`. This will be the name of the layer inside the GeoPackage. Change the *CRS* to the one of the project (EPSG: 32632). Click *OK*.

15. Now add a few other interesting features from OSM (points, lines, and polygons) and add them to the GeoPackage:

- Dams: Key=`waterway`, value=`dam`
- Lakes: Key=`natural`, value=`water`
- Springs: Key=`natural`, value=`spring`

It is important to think about the geometry that you want to download and the order of the layers. Dams look better as lines on top of the lake polygons for example.

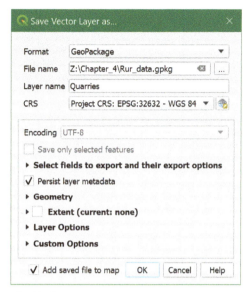

Figure 5.11: Save a Layer to a GeoPackage

16. Now style the layers. Begin with the lakes. You can use the same symbology and label settings used in the Preparing Data from Hard Copy Maps chapter, on page 13. The simplest way to achieve this is to open that map document, right-click on the lakes layer, and choose *Styles | Copy Style | All Style Categories* from the context menu. Then bring up the current map document and again right-click on the lakes layer and choose *Styles | Paste Style | All Style Categories* from the context menu. Switch to the label *Rendering* tab ✐ and click *Only draw labels which fit completely within feature*.

Another option is to open *Layer Properties* for the lakes layer in the map document from the "Preparing Data from Hard Copy Maps" chapter. Switch to the *Symbology* tab and expand the *Layer Rendering* section. Click the *Style* menu and choose *Save style*. In the *Save Layer Style* window save the style as lakes.qml in the exercise folder. Switch to the current project. Open *Layer Properties* for the lakes layer and in the *Layer Rendering* section, click the *Style* menu and choose *Load style*. Choose the lakes.qml file just saved. This is a good option if you will be reusing a style repeatedly.

17. Next you will work with the dam lines. Give them a *Color* of black and a *Stroke width* of 0.86 mm.

18. Click the *Add symbol layer* button 🞡. Select the *Simple line* component and choose a *Symbol layer type* of *Marker line*. Select the *Simple marker* component and choose the vertical line symbol from the choices displayed below. Increase the *Stroke width* to 0.2 mm and the *Size* to 3 mm. Your dams symbol preview should now look like this ┣┼┼┼┼┼┼┫ .

19. Finally, you will style the spring points. Make the layer the target layer in the *Layer Styling Panel*. Select the *Simple Marker* component. Choose a *Symbol layer type* of *SVG Marker*. Select the *symbol* folder and find the *blue-marker.svg* 📍 . Increase the *Size* (width and height) to 6 mm each.

20. Save all temporary scratch layers to the GeoPackage. Load the layers from the GeoPackage and copy the styles. Remove the temporary scratch layers from the *Layers* panel. Save the result

as a new project in the `Rur_data.gpkg` GeoPackage.

6. Calculating Percentage of Land Cover per Subcatchment

6.1 Introduction

For studies on catchment hydrology it is often important to know the percentage of land cover per subcatchment.

After this chapter you will be able to:

- add unique values to vector layers
- calculate areas of polygons in attribute tables
- style a vector layer using a `.qml` file
- *clip, reproject,* and *export* vector layers to another format
- use *conditions* in the *field calculator*
- apply vector *geoprocessing tools* such as *dissolve, intersection,* and *buffer*
- correct geometry errors
- create pie charts using the *Data Plotly* plugin

In this chapter we'll use:

- `catchpolygons.shp`: shapefile with subcatchments
- CORINE 2018 Land cover

The first file has been created by delineating subcatchments within the Rur using the methods described in the Stream and Catchment Delineation chapter, on page 75.

The data can be downloaded from `http://locatepress.com/qgis_hydrological`.

Figure 6.1, on the next page shows the workflow for this chapter.

> For this chapter the videos from the *QGISHydro Chapter 6* playlist at YouTube channel http://www.youtube.com/c/hansvanderkwast provide the theoretical background, results of the steps and additional materials.

6.2 Preparing the Subcatchment Layer

1. Start QGIS Desktop with a new project.

2. Download the `catchpolygons.shp` shapefile and add it to the empty QGIS project.

3. Open the attribute table of the `catchpolygons` layer.

This layer was calculated using the catchment delineation approach from the Stream and Catchment Delineation chapter, on page 75. The result is that each subcatchment has a value of 100. For our purpose, however, each subcatchment needs a unique id.

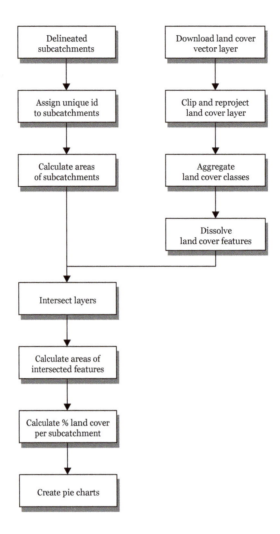

Figure 6.1: Workflow for Calculating the Percentage of Land Cover in Each Subcatchment

4. Toggle to editing mode using the ✏ button.

5. In the *Field Calculator* section just below the toolbar, formulate the equation DN = $id and click *Update All* (figure 6.2, on the next page). Click the 💾 button to save the edits.

The $id function assigns the unique feature id to each feature in the attribute table.

Now we need to add an attribute with the surface area for each subcatchment polygon.

6. Click 🔢 to open the *Field Calculator* dialog.

7. In the *Field Calculator* dialog, create a new field with the *Output field name* CatchArea. Indicate that the *Output field type* is Decimal number (real) with an *Output field length* of 10 and a *Precision* of 2 (figure 6.3, on the facing page).

8. Under the *Expression* tab, type $area. Click *OK* to create the new field with the subcatchment

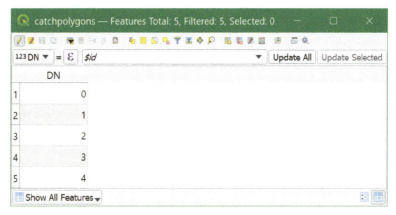

Figure 6.2: Number Features using the Unique Feature ID

areas.

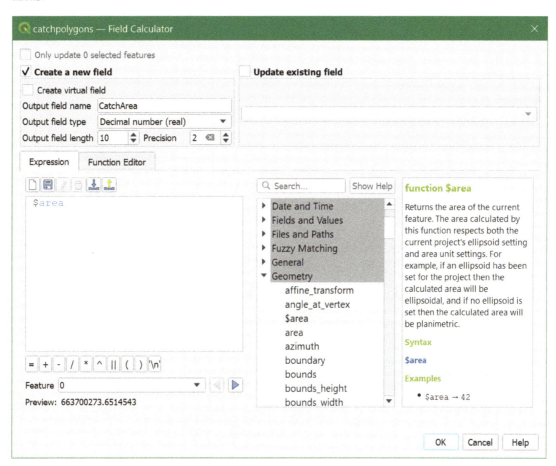

Figure 6.3: Add a Field with Subcatchment Area

The attribute table now has the unique *ids* and surface areas for each subcatchment in square meters (figure 6.4, on the next page).

9. Click the ✎ to save the edits and toggle editing off.

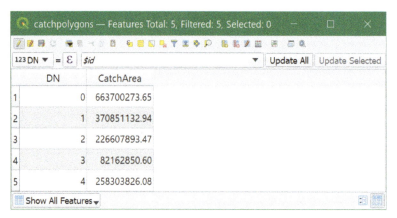

Figure 6.4: Attribute Table with Unique ID's and Areas for each Subcatchment

6.3   Preparing the Land Cover Data

Downloading CORINE 2018 Data

In the previous chapter we used a WMTS layer for visualizing land cover. For calculations, however, we can't use the WMTS or WMS rendered pictures—we need access to the data. For this chapter we will download the CORINE 2018 open data from the Copernicus Land Monitoring Service website.

1. Open a browser and go to https://land.copernicus.eu/pan-european/corine-land-cover.

2. Click on CLC 2018.

You'll see an online map with a preview of the layer.

3. Click the *Download* tab.

4. For downloading data you need an account. Click the link to register.

Once logged in you can choose from different data formats (figure 6.5, on the facing page).

Here we'll use the GeoPackage format, because we need vector data.

5. Check the box before the Corine Land Cover - GeoPackage format.

6. Click the *Download* button.

> The file is 3.5 GB. If you have low bandwidth, you can also download a subset of the dataset from the course materials at http://locatepress.com/qgis_hydrological.

7. Unzip the contents to a new folder that you will use for this exercise.

8. Go to the DATA folder and expand the contents of U2018_CLC2018_V2020_20u1.gpkg.

9. Drag the GeoPackage layer U2018_CLC2018_V2020_20u1 to the map canvas.

10. Click *OK* to keep the default transformation setting.

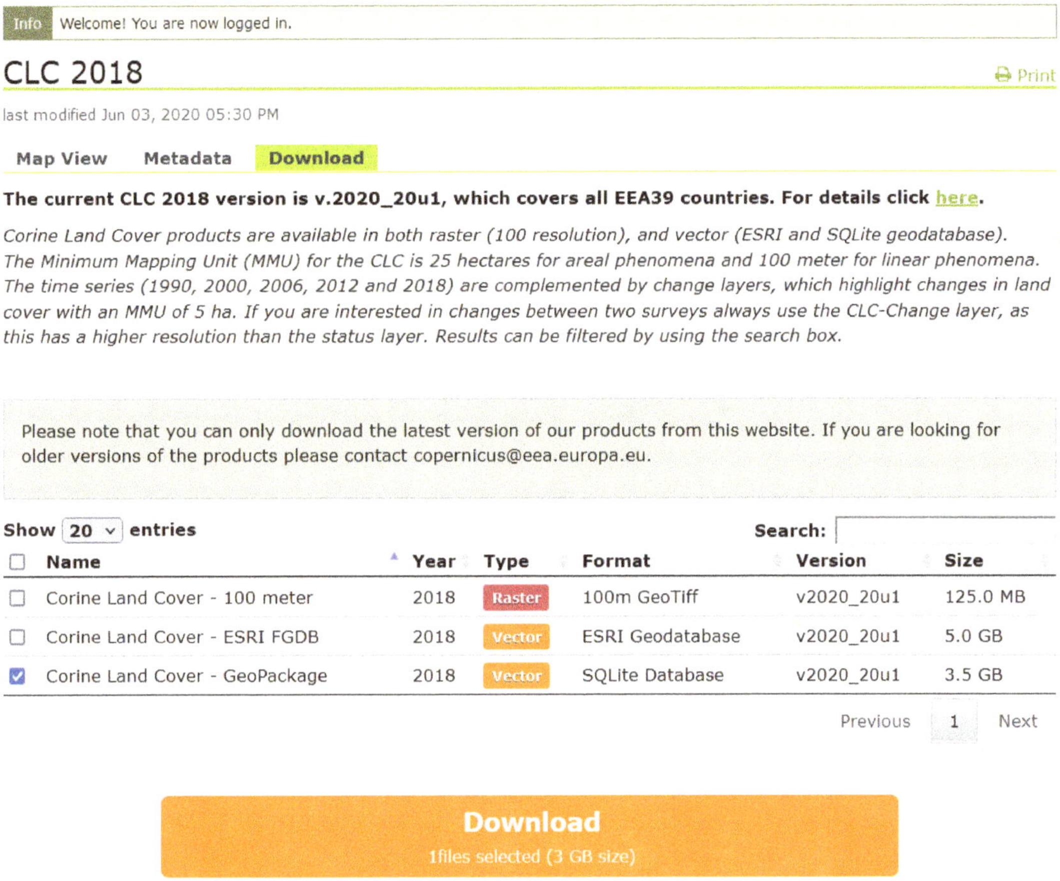

Info	Welcome! You are now logged in.

CLC 2018

🖶 Print

last modified Jun 03, 2020 05:30 PM

Map View   Metadata   **Download**

The current CLC 2018 version is v.2020_20u1, which covers all EEA39 countries. For details click here.

Corine Land Cover products are available in both raster (100 resolution), and vector (ESRI and SQLite geodatabase). The Minimum Mapping Unit (MMU) for the CLC is 25 hectares for areal phenomena and 100 meter for linear phenomena. The time series (1990, 2000, 2006, 2012 and 2018) are complemented by change layers, which highlight changes in land cover with an MMU of 5 ha. If you are interested in changes between two surveys always use the CLC-Change layer, as this has a higher resolution than the status layer. Results can be filtered by using the search box.

Please note that you can only download the latest version of our products from this website. If you are looking for older versions of the products please contact copernicus@eea.europa.eu.

Show 20 ˅ entries   Search: [   ]

Name	Year	Type	Format	Version	Size
☐ Corine Land Cover - 100 meter	2018	Raster	100m GeoTiff	v2020_20u1	125.0 MB
☐ Corine Land Cover - ESRI FGDB	2018	Vector	ESRI Geodatabase	v2020_20u1	5.0 GB
☑ Corine Land Cover - GeoPackage	2018	Vector	SQLite Database	v2020_20u1	3.5 GB

Previous   1   Next

Download
1 files selected (3 GB size)

Figure 6.5: Download CORINE Land Cover as GeoPackage

11. Make the polygons of the `catchpolygons` layer transparent with a black outline and stroke width of 0.66 mm and make sure that it is on top of the land cover layer.

Style the CORINE 2018 Land Cover Map

Now we're going to style the CORINE 2018 land cover map. In the downloaded dataset you will find a `Legend` folder. The folder contains legend files in different formats. Here we'll use the `clc_legend.qml` file, a QGIS layer style file.

12. Right-click on the land cover layer and choose `Properties`. Go to the *Symbology* tab.

13. At the bottom of the dialog, go to `Style | Load Style` (figure 6.6, on the next page).

14. Now you are in the *Database Styles Manager* dialog. Browse to the `Legend\clc_legend.qml` file and click *Load style* (figure 6.7, on the following page).

The legend is now applied to the *Code_18* field of the GeoPackage layer, which contains the CORINE Level 3 land cover classes (figure 6.8, on page 137).

15. Click *OK* to accept and return to the map canvas.

Figure 6.6: Load Style

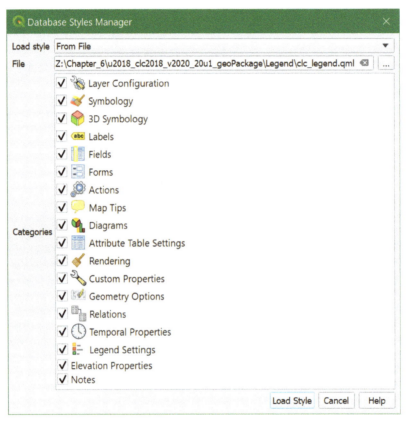

Figure 6.7: Database Styles Manager

Clip and Reproject the Land Cover Map

The land cover layer is still too large (even if you downloaded the provided subset), so we are going to prepare a subset that only covers the subcatchments. In addition, we are going to reproject the land cover layer (EPSG:3035) to the same projection as the catchpolygons layer (EPSG:32632). Both can be done with one tool.

16. Right-click on the U2018_CLC2018_V2020_20u1 layer and choose Export | Save Features As....

17. Choose as a *Format GeoPackage*, and save the U2018_CL2018_V2020_20u1.gpkg GeoPackage with the file name Corine2018_repr, and for *CRS* choose the CRS of the project (EPSG:32632).

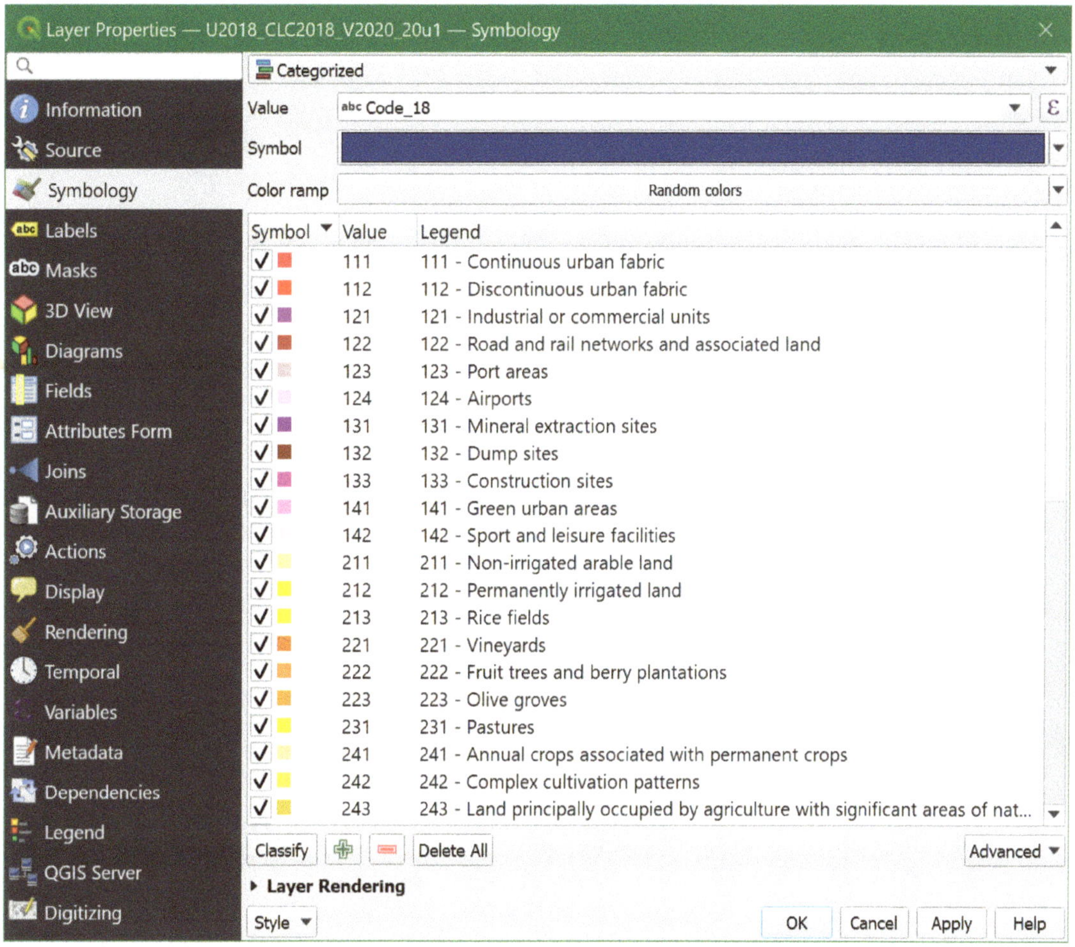

Figure 6.8: CORINE Layer Style from QML File

18. Check the box before *Extent* and click *Calculate from Layer* and choose catchpolygons. Click *OK* (figure 6.9, on the following page).

19. Copy the style from the U2018_CLC2018_V2020_20u1 layer to the Corine2018_repr layer and remove the U2018_CLC2018_V2020_20u1 layer.

The result should look like figure 6.10, on the next page.

Aggregate Land Cover Classes

The CORINE data that we have is termed level 3 data. In the attribute table, the code_18 field has values for level 3, indicated by 3 digits. In this section we are going to aggregate the detailed level 3 classification to level 1. Therefore, we need to create a new field with only the first digit for each feature.

20. Open the attribute table of the Corine2018_repr layer and toggle to editing mode.

21. Click 📅 to open the *Field Calculator* dialog.

22. Add a new field with the name Level1, choose for *Output field type* Whole number (integer).

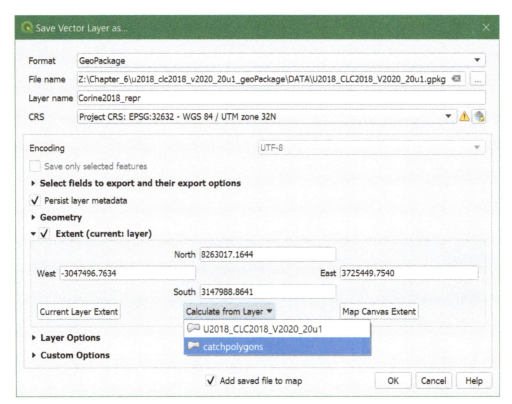

Figure 6.9: Export Layer and Change Projection and Extent

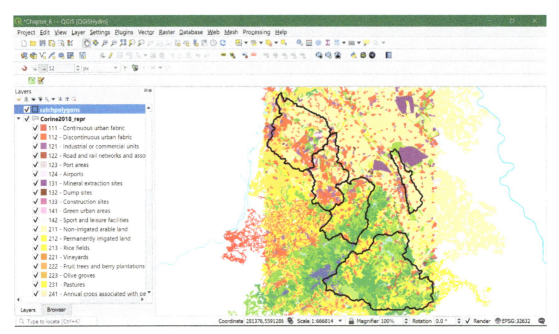

Figure 6.10: Result of Export of the CORINE Data

We are going to write an expression that recodes all level 3 classes to level 1.

23. Write the following expression (figure 6.11, on the facing page): to_int(left("Code_18" ,1)).

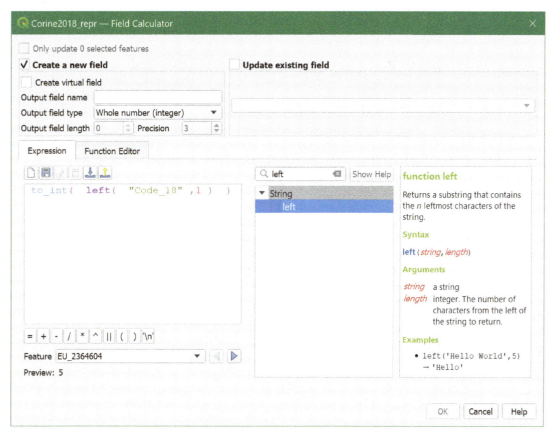

Figure 6.11: Add New Field with CORINE Level 1

The `left` function returns a string with the *n* leftmost characters. Here we need the first character only, so we choose 1. We use the `to_int` function to convert the string to integer.

24. Click *OK* to create the *Level1* field.

The attribute table should now look like figure 6.12.

	OBJECTID	Code_18	Remark	Area_Ha	ID	fid	Level1
1	2364604	511	NULL	84117.024210...	EU_2364604	1	5
2	541777	112	NULL	62.575288924...	EU_541777	2	1
3	542027	112	NULL	489.85208092...	EU_542027	3	1
4	541981	112	NULL	69.123760165...	EU_541981	4	1
5	554446	211	NULL	137.60591445...	EU_554446	5	2

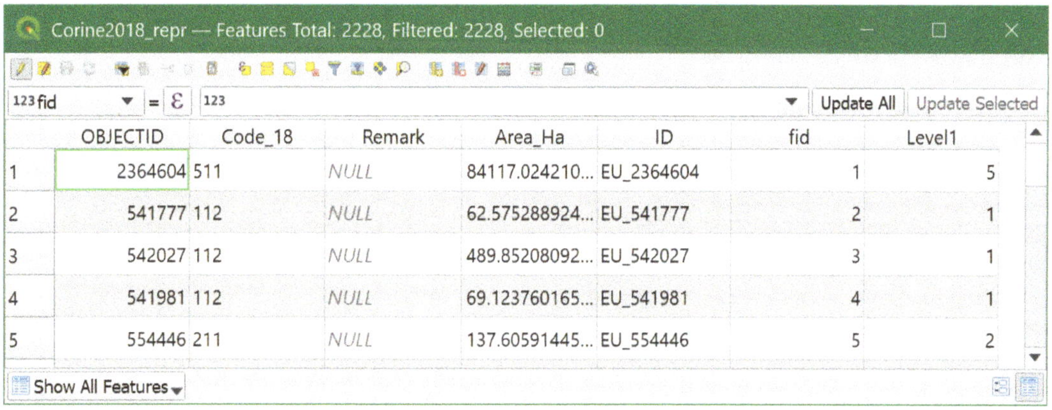

Figure 6.12: Result of the Recoding of CORINE to Level 1

25. Toggle off the editing mode and save the edits.

Dissolve Land Cover Features

Now there are many contiguous polygons with the same level 1 class. These have to be merged into one feature. This is done with the *dissolve* operation.

26. In the main menu, go to Vector | Geoprocessing tools | Dissolve.

27. In the *Dissolve* dialog, choose Corine2018_repr as *Input layer* and select the *level1* field as the *Dissolve field(s)*. Save the file in our GeoPackage as Corine2018_dissolved and click *Run* (figure 6.13).

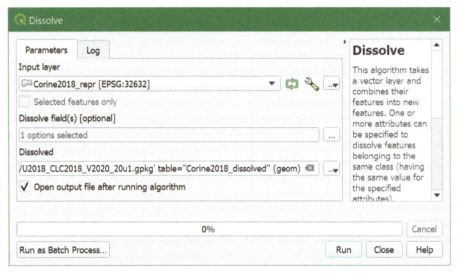

Figure 6.13: Dissolve CORINE Level 1 Features

28. Click *Close* to return to the map canvas when the calculation is complete.

29. Open the attribute table of Corine2018_dissolved to check the result. It should now show only 5 features with level 1 classes 1 to 5 (figure 6.14). Note that you can click on the field name to sort the features.

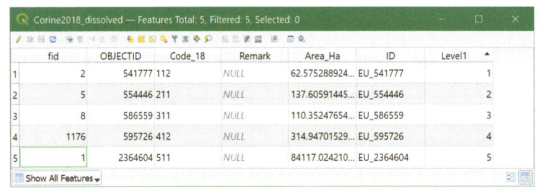

Figure 6.14: Attribute Table CORINE 2018 after Dissolving Level 1 Features

30. Remove Corine2018_repr from the *Layers Panel* and style the Corine2018_dissolved layer with colors based on the original CORINE level 3 legend (figure 6.15, on the facing page). Note that the Level 1 legend is not provided, so you need to choose the colors in the *Layer Styling* panel.

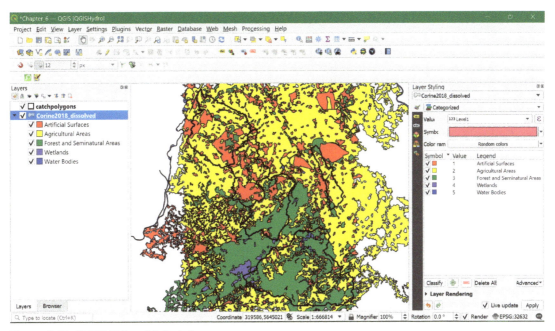

Figure 6.15: CORINE Level 1 Styled

6.4 Intersect the Land Cover Layer with the Subcatchment Layer

Now that we have prepared the land cover and subcatchment polygon layers, the next step is to *intersect* the two layers to add the borders of the subcatchments to the land cover map.

1. In the main menu, go to Vector | Geoprocessing tools | Intersection (figure 6.16).

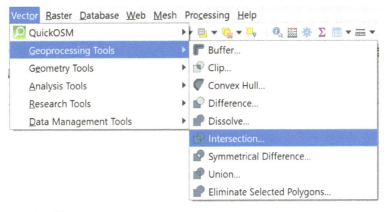

Figure 6.16: Intersection Menu

2. Choose the Corine2018_dissolved layer as *Input layer*, choose catchpolygons as *Overlay layer*. Choose corine_catch_intersected as the layer name in our GeoPackage and click *Run* (figure 6.17, on the following page).

You'll see the following error (figure 6.18, on page 143). This is related to geometry errors that can occur in the catchment delineation process where we polygonize a raster.

3. Click *Close* to close the dialog.

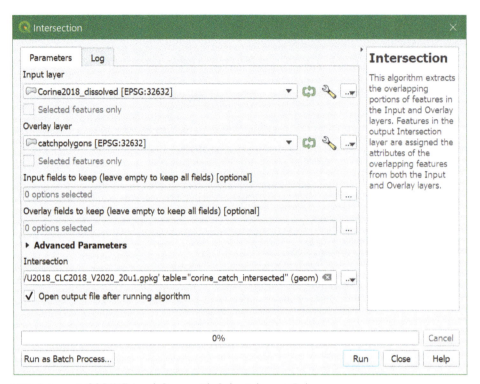

Figure 6.17: Intersect CORINE Land Cover with Subcatchment Polygons

The error can be corrected.

4. In the *Processing Toolbox*, go to `Vector geometry` | `Fix geometries`.

5. In the *Fix Geometries* dialog, choose `catchpolygons` as *Input layer* and save the *Fixed geometries* to the GeoPackage with the name `catchpoly_fixed` and click *Run* (figure 6.19, on page 144).

6. Click *Close* after the processing is done.

7. Copy the style from `catchpolygons` layer to the `catchpoly_fixed` layer and remove the `catchpolygons` layer.

8. Repeat steps 1 and 2 and call the output filename of the intersection layer `corine_catch_intersected_cor.shp`.

9. Style the `corine_catch_intersected_cor` layer by copying the style from the previously styled `Corine2018_dissolved` layer. Check the result of the intersection.

10. Remove the `Corine2018_dissolved` layer from the *Layers panel*.

6.5   Calculate Land Cover Class Area per Subcatchment

Now we need to calculate the CORINE Level 1 class area per subcatchment.

1. Open the attribute table of the `corine_catch_intersected_cor` layer.

In the attribute table, we can find the *CatchArea* field with the area of each subcatchment,

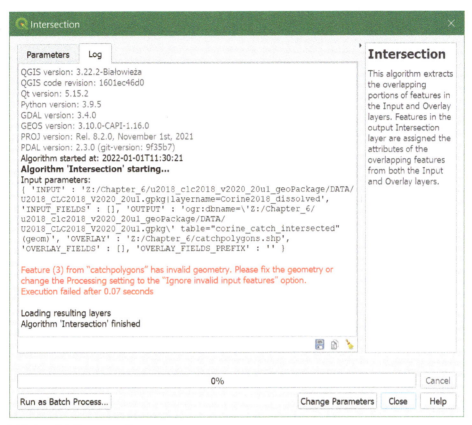

Figure 6.18: Invalid Geometry Error

the *DN* field with the unique ID for each subcatchment, and the *Level1* field with the Level 1 CORINE class for each feature. We will add a new field and calculate the area of each feature which corresponds with the area of each Level 1 class in a subcatchment in a similar way as we have previously done for the subcatchment areas.

2. Use the *Field Calculator* as we did before to create a new field with the name `ClassArea` with *Type* as `Decimal number (real)`. Under the *Expression* tab, type `$area` and click *OK*.

3. Now go back to the *Field Calculator* and create a new field for the percentage of each CORINE Level 1 class in each subcatchment. Call the field `Percentage` with *Type* as `Decimal number (real)`. Under the *Expression* tab type (`"ClassArea" / "CatchArea" ) * 100`. Click *OK* to assign the percentage to each feature.

4. To complete the attribute table, add a field with the Level 1 class names as text. Call it `Landcover`. You can use the `CASE...WHEN` function for that again (figure 6.20, on the next page).

5. Toggle editing off and save the edits.

The attribute table should now look like figure 6.21, on page 145.

6.6   Create Pie Charts Using the Data Plotly Plugin

Now that we have the percentage of land cover for each subcatchment, we can create pie charts. One way is to copy the columns and paste it into spreadsheet software for further processing.

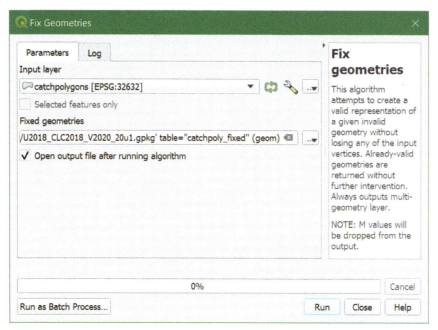

Figure 6.19: Fix Geometries Dialog

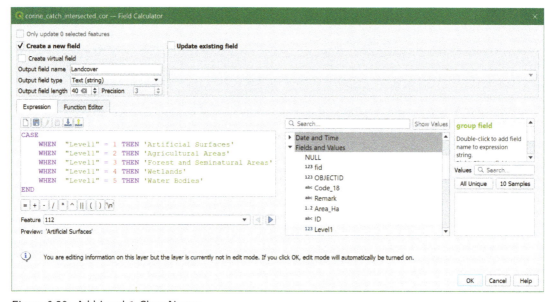

Figure 6.20: Add Level 1 Class Names

Another way is to use the *Data Plotly* plugin, which we will use here.

For the legend of the pie chart, we need to store the colors in the attribute table in hexadecimal codes.

1. Install the *Color to Attribute* plugin from the *Plugins Manager*.

2. Click the *color to attribute* 👥 button in the toolbar.

3. In the *ColorToAttribute* dialog, choose for *Layer* the `corine_catch_intersected_cor` layer

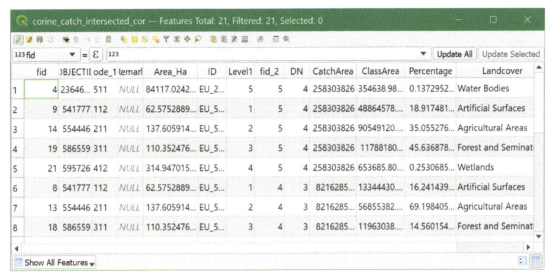

Figure 6.21: Attribute Table after Calculating the Corine Level 1 Class Area, Percentage, and Assigning Class Names

and keep `Create New Attribute` under *Color Attribute*. Type `hex` at the *Attribute name (new)* and click *OK*.

4. Open the attribute table of `corine_catch_intersected_cor` and check the result.

5. Install the *Data Plotly* plugin from the *Plugins Manager*.

We are going to make a pie chart for each subcatchment. We have previously assigned a unique ID for each subcatchment in the `DN` field.

6. Click on the DataPlotly button ⌇ to open the *DataPlotly* panel.

7. For *Plot Type* choose `Pie Chart`. Under *Plot Parameters* choose the `corine_catch_intersected_cor` layer.

8. Click the *Data defined override* ⊟ button at *Feature subset* and choose `Edit....`

9. In the *Expression String Builder* dialog, write the expression `"DN" = 0` so only the subcatchment with a DN value of 0 will be used for the pie chart. In a similar way, you can select the other subcatchments to create a pie chart for those later. Click *OK* to return to the *DataPlotly* panel.

10. For *Grouping field* choose `Landcover`, and for *Y Field* `Percentage`.

11. To create the legend with the same colors as the Level 1 data, click on the *Data defined override* ⊟ button next to *Marker color* and choose `Edit`.

In the *Expression String Builder* dialog, you can see the expected format for the colors. We are going to create an expression that links the classes to the hexadecimal color codes that we have stored in the attribute table.

12. Build the expression as given in figure (6.22, on the next page). The expression maps the land cover categories to the hexadecimal color codes that can be interpreted by the *Data Plotly*

plugin to assign the colors. Click *OK* to return to the *DataPlotly* panel.

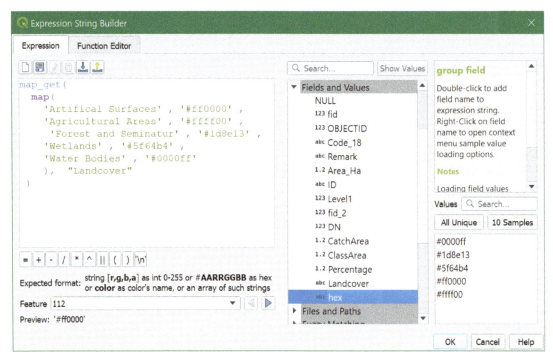

Figure 6.22: Expression to Map Colors to Land-Cover Classes for the Pie Chart Legend

The panel should now look like figure 6.23, on the facing page).

13. Click the ⚙ button to set the layout properties (figure 6.24, on the next page). Add the *Plot title* and check the box for a *Horizontal legend*.

14. Click the *Create Plot* button and check the result.

15. With the *Export to image* button 🖼 you can export the plot to a .png file.

The result should look like figure 6.25, on page 148.

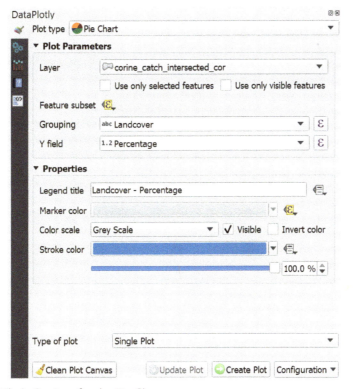

Figure 6.23: Data Plotly Settings for the Pie Chart

Figure 6.24: DataPlotly Legend Settings

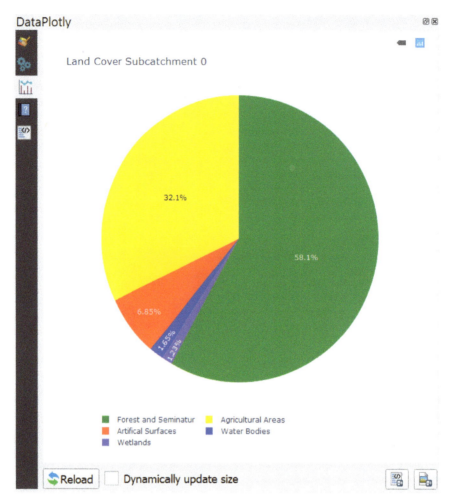

Figure 6.25: Pie Chart Generated with the Data Plotly Plugin

7. Map Design

7.1 Introduction

In this chapter, you will compose a final map showing the result of the catchment delineation analysis. You will start with a basic print layout and enhance it with standard map elements. You will also learn some tricks such as use of variables and expressions to add some automation to the map.

After this chapter you will be able to:

- craft a basic map layout
- add standard map elements
- add *legend patches*
- use a *variable* to add your name to the map
- use *dynamic text* and expressions to automate descriptive text
- work with *map themes*
- export the map

> For this chapter, the videos from the *QGISHydro Chapter 7* playlist at YouTube channel http://www.youtube.com/c/hansvanderkwast provide the theoretical background, results of the steps, and additional materials.

7.2 Map Design Considerations - Adding More Data

In this chapter, you will begin with the map document you saved at the conclusion of the Stream and Catchment Delineation chapter, on page 75. For a basemap, you can choose either the OSM Standard basemap (via the QuickMapServices plugin), or the color hillshade basemap you created in the Stream and Catchment Delineation chapter. The examples that follow will use the color hillshade basemap (figure 7.1, on the following page).

While the data from the analysis are nicely styled, and the study area is highlighted using the *Inverted Polygon Shapeburst Fill* technique, it will be helpful to provide a little context for the map reader. While you have become familiar with the data during the analysis, a map reader needs some additional data to help understand the size and location of the catchment. When crafting a map, it is very important to think about what information needs to be highlighted, who the audience will be, and how will the map be delivered.

For example, you will need to make very different cartographic decisions between a poster, an A4 (or letter), or even smaller peer-reviewed journal sheet sizes. Obviously, the larger the sheet size, the larger the map scale, and the more detail you can include. Also, think about whether it is a technical audience familiar with the data versus the general public. If it is a technical audience, then your work is a little easier. Maps for the general public require more explanation and finesse. Other concerns include whether the map can be produced in color or does it need to be black and white? Color gives you many more options. Producing small

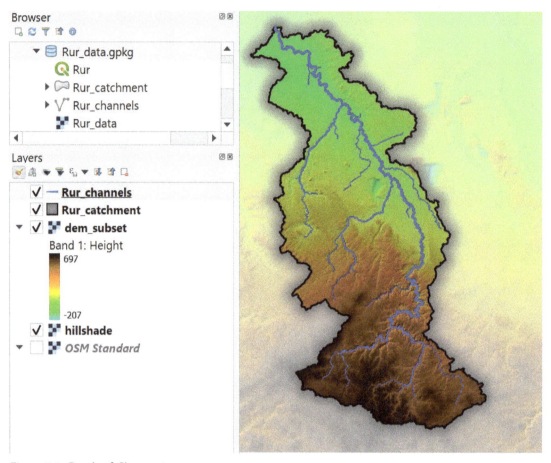

Figure 7.1: Result of Chapter 4

black and white maps for journals can be some of the hardest cartography! Also, think about how the map will be delivered. Will it be printed or distributed digitally?

> You may occasionally need to produce a map for a color blind map reader. Fortunately, QGIS gives you color blind previews. From the *View* menu choose *Preview mode*. Here you will be able to change your screen to simulate: Achromatopsia (Grayscale), Protanopia (No red), Deuteranopia (No Green) and Tritanopia (No blue) along with Monochrome. The ColorBrewer 2.0 website (http://colorbrewer2.org) also gives you options for designing color ramps suitable for a color blind map reader.

For this exercise, you will assume a technical audience. You will produce a map on an A4 sheet in color. This will allow you to either print the map or share it digitally.

1. To begin, you will download cities and towns using the *QuickOSM* plugin. If you need to refresh your memory, refer to the Adding Vector Data from OpenStreetMap section, on page 124 in the Adding Open Data to your Catchment chapter (page 119).

2. Use the *QuickOSM* plugin to download point data with the *Key* of *place*. For that *Key* you will search for and download *Values* of *city* and *town*.

3. You only need to download these data for the extent of the shaded relief.

4. Save these data into the `Rur_data.gpkg` GeoPackage. In this case it is easiest to click the ⊞ and fill in the *Save Scratch Layer* dialog. Name the output layers `Cities` and `Towns`.

7.3   Styling and Labeling Cities and Towns

1. Make sure that the Towns and Cities are at the top of the *Layers* panel.

2. Open the *Layer Styling panel* by clicking the 🖌 button. Set the target layer to *Cities*.

3. They are styled by default with a *Simple marker*.

4. Select the *Marker* component and set the symbol to the *topo pop capital*. This is a symbol that installs with QGIS. It is composed of two different *Simple marker* symbols.

5. Next you will label the cities. Switch to the *Labels* tab (abc) of the *Layer Styling Panel*. Switch from *No Labels* to *Single labels*. Set the *Label with* option to the name field.

6. Change the *Font* to a sans serif font such as *Arial* or *Calibri*. Labels tend to be small. Sans serif fonts are better suited for labels because they are simpler and are easier to read. Reduce the font *Size* to 9 points.

7. To emphasize these cities, change the font *Style* to *Bold*.

Label buffers can be very effective at making labels easier to read, however, you never want the buffer to stand out. They are best when they enhance label readability but aren't initially noticeable.

8. Switch to the *Label buffer* tab **abc** and check the *Draw text buffer* option. Set the *Color* to a light grey (RGB: 191|191|191). Next set the *Blending mode* to *Soft light*.

Now there are subtle buffers that change with the background color.

9. To give more separation between the labels and the feature icon, switch to the *Label placement* ✦ tab and set the *Distance* to 2.5 mm. Also change the *Mode* to *Cartographic*.

Next you will turn your attention to the *Towns*.

10. Set the target layer of the *Layer Styling panel* to *Towns* and switch back to the *Symbology* tab.

11. Select the *Marker* component and set the symbol to *topo pop city*.

12. Switch to the *Labels* tab (abc) of the *Layer Styling Panel*. Switch from *No Labels* to *Single labels*. Set the *Label with* option to the name field.

13. Use the same font you used for *Cities* but reduce the size to 8 points and keep the *Style* of *Regular*.

14. Switch to the *Label buffer* tab **abc** and use the same settings you used for *Cities*.

15. Switch to the *Label placement* ✦ tab and set the *Distance* to 2 mm.

16. Finally, for each point layer, click the *Automated placement settings* ✦ button to open

the *Automated Placement Engine* window. Uncheck the box for *Allow truncated labels on edges of map* option. This will prevent labels from being cut off at the map borders.

17. At this point, your map should resemble figure 7.2.

Figure 7.2: Cities and Towns Styled

7.4 Creating a Catchment Boundary Layer

Currently, the Rur Catchment is styled as an Inverted Polygon Shapeburst Fill. Here you will duplicate the layer to create a version styled as a simple outline.

1. Right-click on the *Rur_Catchment* layer and choose *Duplicate* from the context menu.

2. Turn the duplicated layer on and move it above the Inverted Polygon Shapeburst Fill version.

3. Change the duplicated copy from an *Inverted Polygon* renderer to a *Single symbol*. Select the Shapeburst fill style and click the *Remove Symbol Layer* button to delete it.

Now there is a version of the boundary represented as a simple outline.

7.5   Setting up the Print Layout

Now that you have provided a little context to the analysis results, you will create a new Print Layout and set up the page.

1. First, you will open a New Print Layout. There are three ways to do this.

    - From the menu bar, choose `Project | New Print Layout`
    - Click the *New Print Layout*  button
    - Use the keyboard shortcut `Ctrl + P`.

2. Name the Layout `Rur Catchment` (figure 7.3).

Figure 7.3: Name the Print Layout

3. Click *OK*.

A new Print Layout will open. This is where you craft your map.

The Print Layout is an application window with many tools that allow you to craft a map. For detailed information about the Print Layout, refer to the QGIS manual: `https://loc8.cc/dq3/composer`. The main window of the Print Layout displays the piece of paper upon which the map will be designed.

There are buttons along the left side of the window that allow you to add various map elements: map, scale bar, photo, text, shapes, attribute tables, etc. Note that each item added to the map canvas becomes a graphic object that can be moved, resized, and further manipulated (if selected) by the *Item Properties* tab on the right side of the layout. Across the top are buttons for exporting the composition, navigating within the composition, and some other graphic tools (grouping/ungrouping etc.) as shown in figure 7.4, on the following page.

4. To set the sheet size, right-click on the blank page and choose *Page Properties*. Here you can specify details about the overall composition. QGIS defaults to an A4 sheet size. There is no need to change that here since that is what you will use.

> QGIS Print Layouts allow you to add as many pages to a Layout as you wish for a given map document. These can also be of differing page sizes and orientations. They can even be in a different CRS from the main map canvas.

5. Set the *Orientation* to *Portrait*.

> Note: Clicking the Save button in a Print Layout saves the entire map project. A project can contain multiple Print Layouts.

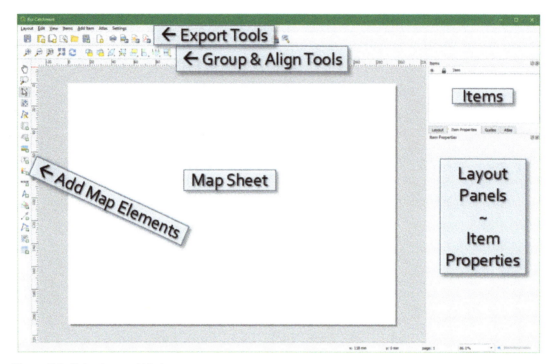

Figure 7.4: Print Layout Window

7.6   Adding the Map

1. Using the *Add new map to layout* 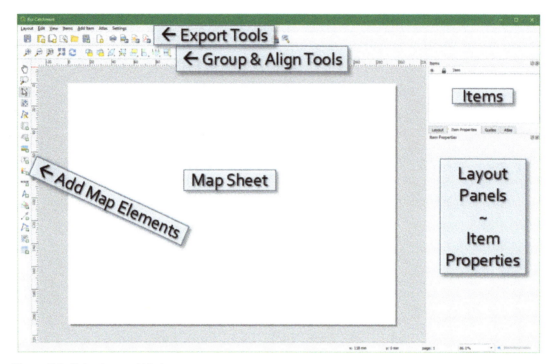 button, drag a box on the map sheet where you'd like the map to go. Remember that you'll need room for a title at the top of the page and a legend at the bottom of the map (figure 7.5, on the next page).

The map object can be moved and resized after its been added by selecting it with the *Select/Move item* tool. You can then use the handles around the perimeter to resize it. Remember when an object is selected, the *Item Properties* tab will show properties specific to that object.

The next step is to set the map extent within your composition.

2. With the map selected click on the *Item Properties* tab. There are a series of buttons across the top of the panel for controlling the map extent.

From left to right these are:

- Update Map Preview
- Set Map Extent to Match Main Canvas Extent
- View Current Map Extent in Main Canvas
- Set Map Scale to Match Main Canvas Scale
- Set Main Canvas to Match Current Map Scale
- Interactively Edit Map Extent
- Labeling Settings

- 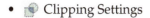 Clipping Settings

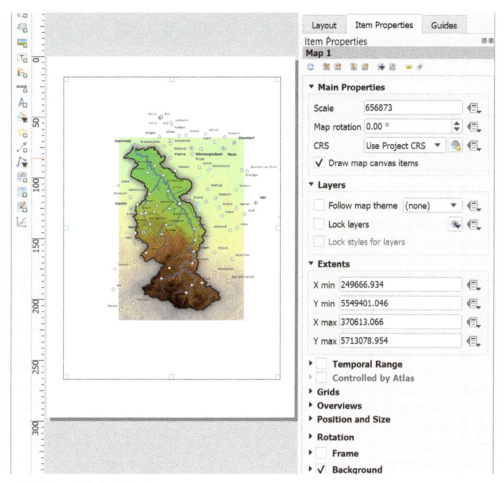

Figure 7.5: Map Added to Print Layout

3. Click the *Set Map Extent to Match Main Canvas Extent* 🗺 button. That will help orient the map on the sheet of paper as it appears in QGIS Desktop. However, if the aspect ratio of your QGIS Desktop map canvas doesn't match that of your print composition this may not give you the desired result.

If you need to make additional adjustments to the scale, you can do so in the *Main properties* section by adjusting the *Scale* value. Map scale is a ratio of Map Distance/Ground Distance. Here the number is roughly 750000 which can be read as a scale of 1:750,000. To zoom out you increase this number—reducing it zooms in. Clicking the *Update Map Preview* 🔄 button forces the map view to refresh.

4. Set the *Scale* value to 395000 and hit the enter key. This should zoom the map to the boundaries of the Rur Catchment.

5. If you need to pan the map, you can use the *Move Item Content* 🔳 button. This allows you to pan the map content in the map frame without changing the scale. It is normal to have to make adjustments to get the map extent just right. Try to make your layout match figure 7.6, on the following page.

6. Scroll down on the *Item Properties* tab and find the *Frame* section. Enable the *Frame* and increase the width to 0.8 mm.

Make sure that there is no gap between the map content and the frame.

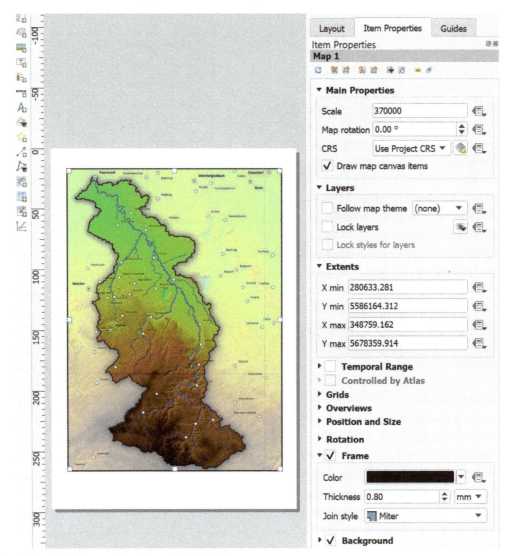

Figure 7.6: Print Layout Map Extent Established

7.7 Adding a Title

Now you will add the title to your map. The purpose of a map title is to quickly convey the content and focus of the map to the reader. It should be concise and prominent.

1. Use the *Add new label* tool to drag a box all the way across the top of the composition above the map object. The text box can be resized after the fact by using the graphic handles.

2. By default, the text box will be populated with the placeholder text *Lorem ipsum*. Using the *Item Properties* tab *Main Properties* section, replace the holder text with the title: `Rur Catchment and Channels`.

3. In the *Appearance* section you can change the font. Click the *Font* button to open the *Text Format* window. Change the font to: Times New Roman, Bold, Size 36. You can use the search box above the font list to search for Times New Roman.

4. Below the font settings are some alignment settings. Set the *Horizontal alignment* to *Center* and *Vertical alignment* to *Middle* (figure 7.7).

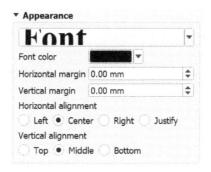

Figure 7.7: Title Font Settings

7.8 Adding a Legend

The purpose of the legend is to identify what symbols and colors on the map represent. Legends are used for data layers that are non-intuitive or require more explanation. For example, a point labeled *Rotterdam* does not need to be included in a legend, it is obvious what it is. However, the Strahler orders do need explanation.

Adding a Legend for Vector Layers

The *Interpolated Line* renderer that we have used to style the Strahler orders was a great way to visualize the Strahler orders, but it does not result in a legend with the Strahler orders as classes of lines with a different thickness. We can create this by using the *Graduated* line renderer.

1. Return to the main QGIS window.

2. Duplicate the Rur_channels layer.

3. Hide the Rur_channels layer and make the Rur_channels copy layer visible.

4. Go to the *Layer styling* panel and make sure Rur_channels copy is active.

5. Change the Single Symbol renderer to Graduated and change the following settings:

- Set the *Value* to ORDER.
- Set the *Symbol* to a simple blue line by changing the *Interpolated Line* to *Simple line* with RGB color 15 | 66 | 220. Set the *Width* units to Millimeters, otherwise the scaling by size will not work properly.
- Set the *Method* to Size.
- Set the *Size from* 0.3 *to* 1.0 mm.
- Set the *Mode* to Natural Breaks (Jenks) with the same number of Strahler Order classes that exist in your data. Review the attribute table if need be. In this example, it is set to 3 classes. Note that you might need to adjust the ranges, based on the amount of Strahler orders that you have. The first range always includes the lower and upper boundary

value, while all the other ranges exclude the lower boundary, but include the upper boundary.

- Set the *Legend* values to integer Strahler orders.

Your *Layer Styling* panel should now look like figure 7.8.

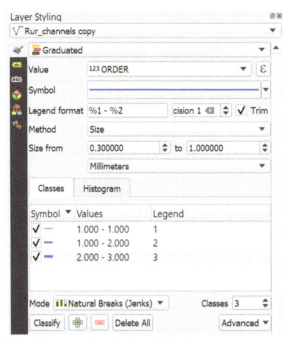

Figure 7.8: Channels Styled by Size using the Graduated Renderer

The result should be similar to the original Rur_channels.

6. Return to the *Print Layout* and use the *Add new legend* tool to drag a box on the lower-right corner beneath the map.

The legend will not fit in this space with the default font but you will change the settings. The only layers needed are the catchment and the channels. The *Item Properties* tab will be used to configure the legend (see figure 7.9, on the next page).

7. Uncheck *Auto update*. This will enable you to modify the legend, however, updates to the map will no longer be reflected in the legend unless you re-enable *Auto update*.

8. Select the *Cities* layer and click the *Remove item* button to remove it. Repeat to remove all the layers except the *Rur_channels copy* and *Rur_catchment copy*.

9. Use the *Move Up* and *Move Down* buttons to move the *Rur_catchment copy* to the top of the legend.

10. Layers in the legend should not have original file names. They should be descriptive names that will be clear to the map reader. Now you will work on renaming the two layers. Select the *Rur_channels copy* layer name and click the *Edit* button. Change it so it reads *Strahler Stream Orders*.

11. Rename *Rur_catchment copy* to *Rur Catchment Boundary*.

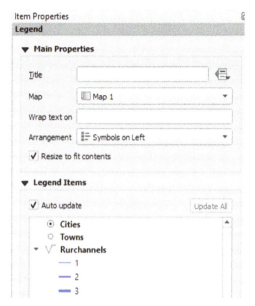

Figure 7.9: Legend Properties

12. Scroll down in the *Item Properties* panel to find the *Columns* section. Expand it. Set the *Count* to 2.

Importing Legend Patch Shapes

Now you have completed a simple legend for the vector layers. Next you will learn to make a more intuitive legend by using *Legend Patch Shapes*. To begin, you'll download a set from the QGIS Styles Repository.

13. Return to the main QGIS application.

14. Click on the *Style Manager* button.

In the *Style Manager*, you can manage styles for marker, line and polygon symbols, color ramps, text, labels, legend patch shapes, and 3D symbols (figure 7.10).

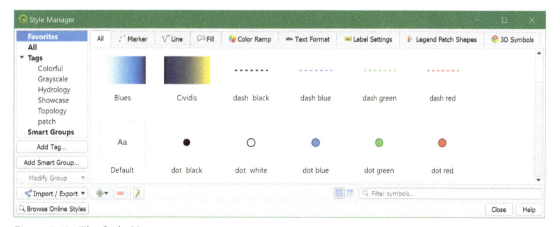

Figure 7.10: The Style Manager

In the Style Manager, you can import items from XML files or from URLs. You can also download styles from the online QGIS Style Repository by clicking *Browse Online Styles*. There you can also upload your own styles and share them with other users.

There is a good online resource for legend patches in the QGIS Style Repository.

15. Click *Browse Online Styles*.

16. Download the *Basic Legend Patches Set* by Klas Karlsson and extract the XML file in the zip file.

17. Return to the *QGIS Style Manager* and select the *Legend Patch Shapes* tab. It is empty by default, but you will now import the legend patch shapes in the Klas Karlsson XML file.

18. Click *Import/Export | Import Item(s)*.

19. Set *Import from to File*.

20. Browse to the patch.xml file.

21. Click *Do not import embedded tags*.

22. Click *Select All* and *Import* (see figure 7.11).

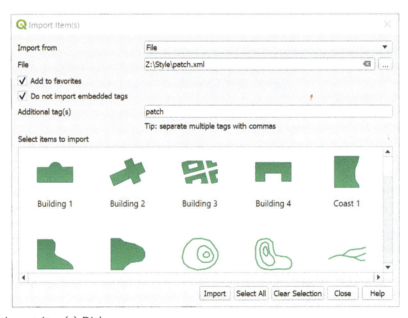

Figure 7.11: Import Item(s) Dialog

You will now see more than a dozen new legend patches to choose from.

23. Close the *Style Manager*.

24. Open your *Print Composition*.

25. Select the *Legend*.

26. Double-click on *Strahler Stream Orders* in the *Item Properties* tab. Under the *Patch* section

click on the *Shape* symbol.

27. Change from `Favorites` to `All Legend Patch Shapes` in the drop-down menu.

28. Choose the *River 1* patch symbol (figure 7.12).

29. Click the *Go back* ◀ button and you will have the Strahler orders in the legend with nice legend patches in a river style.

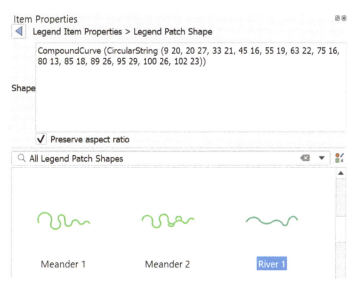

Figure 7.12: Select the River 1 Patch Shape

Creating Custom Legend Patch Shapes

The rectangle representing the catchment boundary in the legend can also be improved. We are now going to add the shape of the Rur catchment as a custom legend patch.

30. Return to the main QGIS window.

31. Select `Rur_catchment` copy in the *Layers* panel and then use the *Select Feature(s)* button to select the catchment boundary polygon in the map canvas.

32. Toggle on the editing mode for the `Rur_catchment` copy layer.

33. Click the *Copy features* button in the *Digtizing toolbar*.

34. Open the *Style Manager* by clicking the button in the toolbar.

35. Switch to the *Legend Patch Shapes* tab and click *All* in the left side panel to show all patches.

36. Click the *Add item* button and choose *Fill Legend Patch Shape...* (figure 7.13, on the next page).

37. In the *New Legend Patch Shape* dialog (figure 7.14, on the following page), delete the default code in the *Shape* box and use `Ctrl + V` to paste the catchment polygon geometry here in the *GeoJSON* format.

Figure 7.13: Add Fill Legend Patch Shape Item to the Style Manager

38. Scroll to the top and delete the wkt_geom fid DN line.

39. Scroll to the bottom and delete everything after the brackets.

40. Click *Preserve Aspect Ratio*

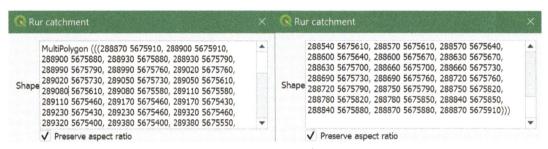

Figure 7.14: New Legend Patch Shape Dialog

41. Click *OK*.

42. In the *Save New Legend Patch Shape* dialog (figure 7.15), you can *Name* the symbol as Rur catchment and add a Hydrology tag. Check the box to *Add to favorites*. Click *Save*.

Figure 7.15: Save New Legend Patch Shape Dialog

You can now see the new legend patch shape under *Favorites* in the *Style Manager*.

43. Click *Close*

44. Toggle off editing and deselect the catchment boundary with the *Deselect Features from All Layers* button.

45. Return to the *Print Layout*.

46. Change the *Rur Catchment Boundary* symbol to the *Rur catchment* shape in the same way as we have added the river legend patch shape before.

47. Increase the *Height* of the patch to 20 mm and add a return before *Boundary* to move it to the next line.

48. You can remove the *Background* of the legend if needed.

The result should look now like figure 7.16.

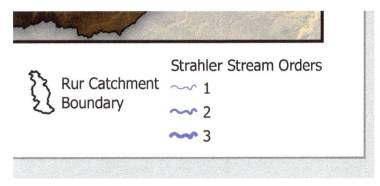

Figure 7.16: Legend Patch Shapes Configured

Adding the Elevation Legend

We also need to add a legend for the elevation. With the current layout, we have room along the lower left corner of the map where this second legend will nicely fit.

49. Use the *Add new legend* tool and drag a rectangle in the lower-left corner.

50. In the *Item Properties* tab, uncheck *Auto update*. Remove all layers except dem_subset.

51. Select *Band 1: Height* and remove it by clicking the *Remove item* button.

52. Double-click on *dem_subset* and replace the *Label* with Elevation. Under *Patch*, set the *Width* to 5 mm and the *Height* to 50 mm (figure 7.17).

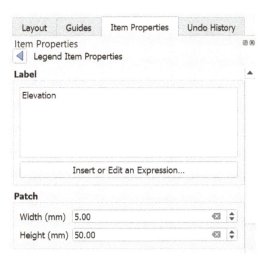

Figure 7.17: Elevation Legend Label Settings

53. Click the *Go back* button to return to the legend settings.

54. Double-click on the color ramp.

55. At *Suffix*, type a space and m for the units.

56. Under *Layout*, change the *Direction* to Minimum on Top. We do that here, because it then corresponds with the elevation gradient on the map (figure 7.18).

By default, the minimum and maximum values used for the styling are added as labels. In our case, those are the values of the full extent of the dem_subset layer. If you want to use the values for the elevation within the catchment, you need to use a DEM that is clipped to the boundary of the catchment and type its minimum and maximum value here.

Figure 7.18: Elevation Legend Layout Settings

57. Click the *Go back* ◀ button to return to the legend settings.

58. Uncheck the box for *Background* to make the background transparent.

Now the elevation legend looks like figure 7.19, on the next page.

7.9   Adding a Scale Bar

Scale bars give the map reader a way to approximate distances on the map. There are two types: graphic scale bar or scale text. Here you will learn how to add a graphic scale bar.

1. Click on the *Add new scalebar* button.

2. Click to the left of the legend to add the scalebar to the map.

3. On the *Item Properties* tab, in the *Units* section, make sure the *Scalebar units* are set to *Kilometers*.

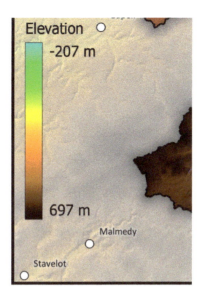

Figure 7.19: Gradient Legend

4. In the *Segments* section, keep the left to 0 and the right to 2.

5. In the *Main Properties* section, change the *Style* to *Stepped Line*. Notice that one of the styles is *Numeric* which would add scale as scale text (*1:100,000*).

6. In the *Display* section, click on *Font* and reduce the size to 9 points.

7. Use the *Select/Move Item* tool to place the scalebar in a good position near the bottom of the white space in the lower left corner. Next, you will be adding some descriptive text and a north arrow just above the scalebar, so leave space for those (figure 7.20).

Figure 7.20: Scale Bar Added

7.10   Adding a North Arrow

Often it is nice to add a north arrow to a map composition to help orient the map reader. One should especially be added if north is not up on the map. Here you will learn how to add this to your map.

1. Click on the *Add north arrow* button.

2. Drag a small box into the empty space just right of the scale bar. A default north arrow graphic will be added.

3. If you want to choose a different north arrow, you can find a series of SVG graphics included with QGIS under *SVG browser* in the *Item Properties* tab. Under *SVG Groups | App Symbols | arrows* you can choose a different one if you'd like.

4. Scroll down to the *Image Rotation* section of *Item Properties*. Note that *Sync with map* is enabled and is set to *Map 1* (figure 7.21).

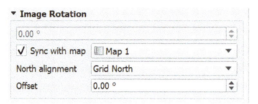

Figure 7.21: North Arrow Image Rotation Settings

5. Scroll up to the *Size and Placement* section and change the *Resize mode* to *Zoom and resize frame*.

6. Resize the north arrow graphic and move as needed so that it is well placed (figure 7.22).

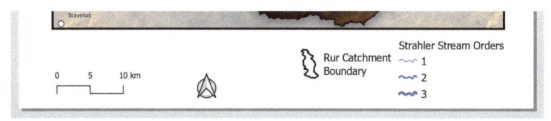

Figure 7.22: North Arrow Placed on Map Composition

7.11   Adding Descriptive Text

It is good practice to include credits for both data sources and cartography on a map. It can also be helpful to include details such as the date.

Next, you will enter some descriptive text that tells the map reader where the data was obtained, who the cartographer was, and the date created. This will be done using the *Add new label* tool, the same tool you used to add the title.

1. Add the label in the space above the scale bar and north arrow. Add the following text in the *Main properties* window of the *Item Properties* tab:

```
Data Sources:  SRTM, Natural Earth and (c) OpenStreetMap Contributors

Cartographer:  <your name>

Created on:  <todays date>
```

You can manually enter the date. However, it is also possible to use an expression to create *Dynamic Text* for the date.

2. Select Dynamic Text | Current Date | Day Month Year. An expression for the current date is entered. This will update automatically each time you open this print layout. This means it is no longer necessary to remember to update this every time you edit the map.

3. This created the following expression for the date: [%format_date(now(), 'dd/MM/yyyy')%]. This uses the format_date() with the now() function to express the current date. The format of the date can be altered by changing the formatting string: 'dd/MM/yyyy'.

4. In the same way, you can add an expression to use the copyright symbol for the credits to OpenStreetMap. In Chapter 2 you learned to use the char function for that. The code for the copyright symbol is 0169.

5. To finish this element, set the font to *size* of 8.

6. Configure the text element, scale bar, and north arrow so that they all fit in the space to the right of the legend. Note that with an element selected, you can also use the arrow keys on your keyboard to nudge it.

> Did you know there is a way to do custom layout checks? An example would be checks to warn a cartographer if desired map elements have been added or properly configured. These checks can help ensure maps meet a set of minimum organizational design criteria. Read more here: http://bit.ly/2G852H8

7.12 Using Variables for Adding Your Name as Author

QGIS allows you to store variables. This means any type of constant such as a unit conversion factor or your name. Variables can be set at several levels: Global, Project, and Layer. Here you will create a new Global variable with your name as cartographer.

1. Bring up the main QGIS desktop application.

2. From the menu bar, choose Settings | Options.

3. Click on the ε Variables section. This is where *Global* variables are found.

4. Click the *Add variable* button.

5. Replace *new_ variable* with *cartographer*.

6. Click in the *Value* cell to the right and type your name (figure 7.23, on the next page). Click *OK*.

7. Bring up the *Print Layout* window again.

8. Select the text you were most recently working with.

9. Highlight the name you entered for cartographer and click the *Insert an Expression* button.

10. Scroll down in the list of functions to the *Variables* section and expand it. Find the variable you just created named *cartographer*. Double-click on it to add it to your expression. When you select it, notice that the value of that variable is displayed in the right-hand pane of the expression window (figure 7.24, on the following page).

While you have the *Insert Expression* window open, notice that there is a *Recent* section. Expand it and you will find the date expression you created. You can use this *Recent* section to bring up recently used expressions and use them without having to recreate them!

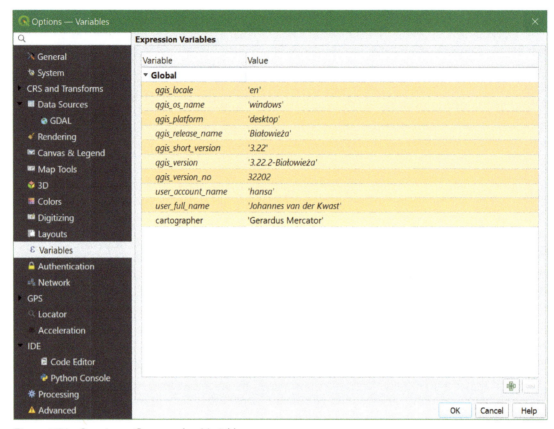

Figure 7.23: Creating a Cartographer Variable

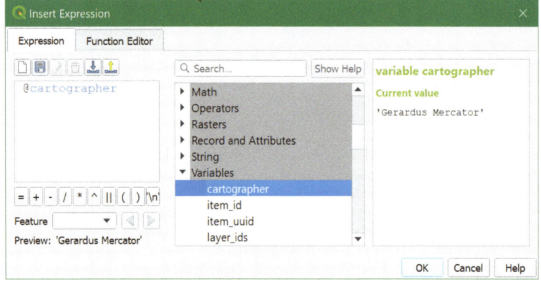

Figure 7.24: Using the Cartographer Variable

11. Click *OK*.

12. The text in the *Items Properties* panel now reads: (figure 7.25).

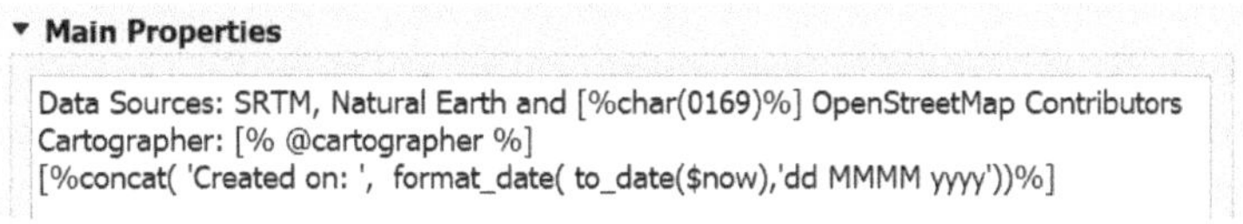

▼ **Main Properties**

Data Sources: SRTM, Natural Earth and [%char(0169)%] OpenStreetMap Contributors
Cartographer: [% @cartographer %]
[%concat('Created on: ', format_date(to_date($now),'dd MMMM yyyy'))%]

Figure 7.25: Text with both Date Expression and the Cartographer Variable

You have now used both an expression and a variable to automate your text. The @cartographer variable will always be available until you delete it. Variables and other expressions can be used throughout the QGIS interface to make your job easier. The date expression will update automatically each time the map is modified.

Your overall map should now resemble figure 7.26, on the following page.

7.13 Setting Up a Map Theme

In these final sections, you will learn how to work with *map themes* to create a locator map. Locator maps are smaller than the main map. They have a smaller scale and provide a broader spatial context, showing where the main body of the map is located.

This will involve:

- Creating a map theme from the current map and connecting the map layout to the theme
- Adding, styling, and labeling a country boundaries layer
- Creating a map theme for the locator map
- Adding a new map to the layout and connecting it to the locator map theme
- Configuring the Overview

1. Return to the QGIS Desktop window. Select all the layers in the *Layers panel* and choose *Group selected* from the context menu. Name the group *Main Map* (figure 7.27, on page 171).

2. Now you will set the current view as a map theme. Locate the *Manage Map Themes* 👁 drop-down menu at the top of the *Layers Panel* and click on it. Choose *Add Theme* and the *Map Themes* window will open. Name the new theme *Main Map* and click *OK*.

3. Return to the print layout window. Select the map object with the *Select/Move Items* tool. On the *Item Properties* tab find the *Layers* section. Click the box for *Follow map theme* and change it from *none* to *Main Map* (figure 7.28, on page 171).

7.14 Setting up the Layers for the Locator Map

Here you will add some data, set it up in a new Layer group and style it. The new Layer group will have two layers: Country boundaries and a copy of the Rur Catchment.

1. From the data included with the book, find the MapComposition.gpkg. Add the CountryBoundaries layer found inside. If it lands within the existing *Main Map* group, right-click on it and choose *Move Out of Group*.

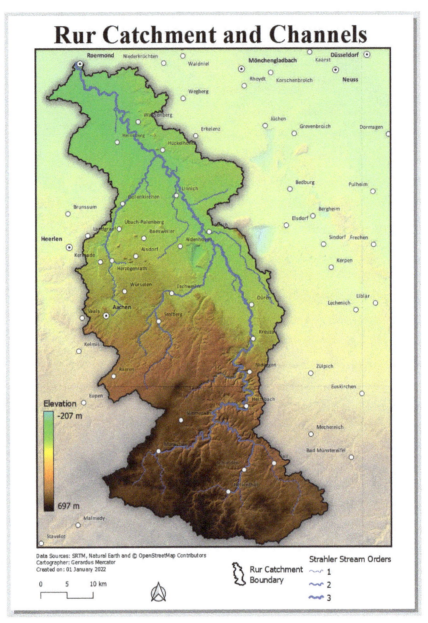

Figure 7.26: The Map Composition at this Point

Figure 7.27: Main Map Layers Grouped

Figure 7.28: Setting the Map Object to Follow a Map Theme

This dataset was downloaded from Natural Earth (https://www.naturalearthdata.com). This is another fantastic public domain resource for global GIS data.

2. Click Ctrl + Shift + H to Hide All Layers. This is a keyboard shortcut that allows you to quickly turn off all the layers in a project. Turn the CountryBoundaries layer back on.

3. Find the copy of the Rur Catchment layer that is styled with a simple outline (not the one styled as an inverted polygon shapeburst fill). Right-click on it and choose *Duplicate*. Then right-click on the copy and choose *Move Out of Group*. You now have the two layers that will participate in the Locator map.

4. Select both these layers and choose *Group selected* from the context menu. Name the group *Locator Map* (figure 7.29, on the next page).

5. Style the CountryBoundaries with a *Fill style* of *No brush*. Keep the default *Stroke color* of black but increase the *Stroke width* to 0.46 mm.

6. Switch to the *Labels* tab ⟨abc⟩ of the *Layer Styling Panel*. Label the CountryBoundaries with the NAME field.

7. Change the *Font* to a sans serif font such as *Arial* or *Calibri*. Reduce the font *Size* to 9 points and change the font *Style* to *Bold*.

8. Now you will style the Rur Catchment for the Locator map. Select the *Simple line* compo-

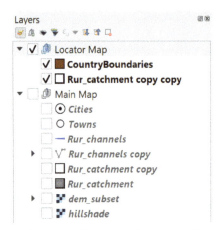

Figure 7.29: Both the Main Map and Locator Map Layer Groups Configured

nent. Change the *Symbol layer type* to *Simple fill*.

9. Change both the *Fill color* and *Stroke color* to RGB: 0|145|255 (figure 7.30).

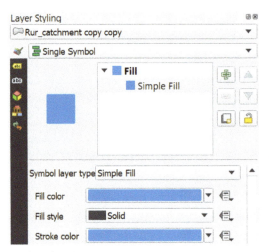

Figure 7.30: Rur Catchment Styling for the Locator Map

10. Now you will set the current view as the Locator map theme. From the *Manage Map Themes* drop-down menu again choose *Add Theme*. In the *Map Themes* window name the new theme *Locator Map* and click *OK*.

11. Return to the print layout window. You will now add a second map to the layout.

12. Using the *Add new map to layout* button, drag a small box on the existing map object along the upper right side, being careful to not cover the Rur Catchment.

13. You will now set the *Map theme* for this second map. On the *Item Properties* tab, find the *Layers* section. Click the box for *Follow map theme* and change it from *none* to *Locator Map* (figure 7.31, on the next page).

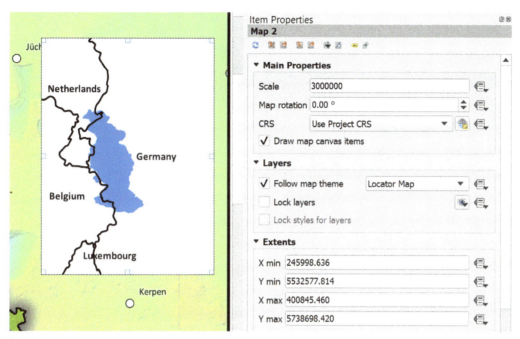

Figure 7.31: Locator Map Added

On the Item Properties tab for map objects, there is also an *Overviews* section which allows you to configure a box indicating for example, the spatial envelope of the main map canvas within the locator map. In this case, you did not use that because including the watershed on the locator is sufficient. However, it is often useful to include *Overviews* with inset and locator maps.

7.15 Final Adjustments

1. Return to the main QGIS Desktop window. Give the labels for the CountryBoundaries a default white *Buffer*. This will help the labels be more readable where they cross national boundaries.

2. Return to the print layout window and click the *Refresh map* button to see the change.

3. Make sure the Locator Map is still selected. From the *Item Properties* window, decrease the scale (zoom out) of the Locator map to about 3000000 so the labels better fit within the national boundaries.

4. Scroll down and activate the *Frame* for the Locator map.

5. It may be that the Locator map is interfering with City and Town labels on the main map. To avoid this conflict, select the Main map object. Near the top of *Item Properties* find the *Labeling Settings* button.

6. This opens the *Label settings* panel. Here you can select map elements as label blocking elements (figure 7.32, on the following page). Click *Map 2*. Now the labels for Cities and Towns on the main map will move to avoid conflicts with the Locator map. This can be done for any the map elements.

Did you know that you can also move labels to custom positions? For example, you may want to shift some country labels in the Locator map. From the main QGIS Desktop window enable the Labels toolbar. Select the layer with the labels you wish to move. Find the button on the toolbar named *Move Label and Diagram*. Click on a label. You will be prompted to identify the unique ID column of the layer. Once that has been done you can click on individual labels and move them to custom locations. On the same toolbar there are also tools for Rotating labels and altering fonts (Change label).

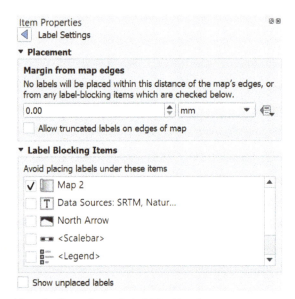

Figure 7.32: The Locator Map Configured as a Label Blocking Item

7.16   Exporting the Map

Congratulations your map is finished! The final step is to export it to a high-resolution image.

1. Click the *Export as image* button.

2. The *Save Layout As* window opens.

   - Navigate to the exercise folder.
   - Choose a *Save as* type such as JPG or PNG.
   - Name the file and click *Save*.
   - The *Image Export Options* window will open. Keep the default 300 dpi setting and click *Save*.

Note that maps can also be exported as a PDF or SVG file.

3. Once the map has finished exporting, a link to the folder will appear across the top of your print layout. You can click on the link to open your systems file browser and see the result (figure 7.33, on the next page).

4. The final map should look like figure 7.34, on page 176.

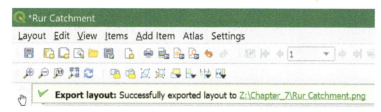

Figure 7.33: Map Export Link

Rur Catchment and Channels

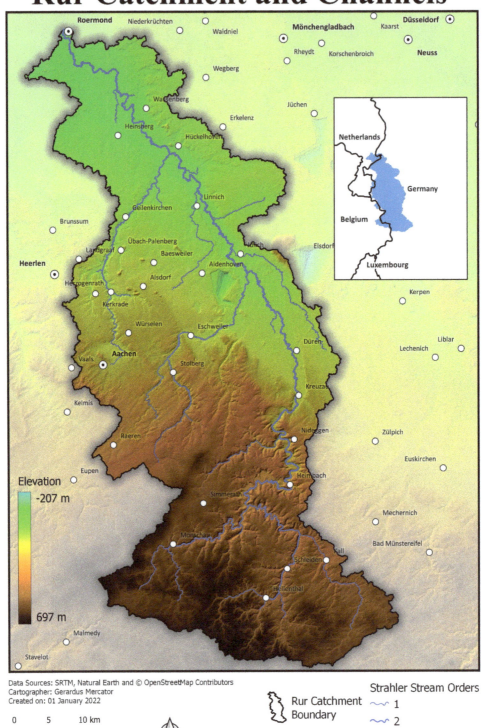

Figure 7.34: Final Map

8. Conclusion

Having reached this chapter, you've made it through the exercises of QGIS for Hydrological Applications. We hope you've enjoyed learning useful features and plugins that QGIS offers for catchment hydrology and water management. Now it is important to apply the skills to your own cases and add different flavors to the recipes presented in this book.

It was hard to make a selection of exercises for the book. QGIS has many more useful features and plugins to explore. Besides QGIS, there are many other FOSS4G tools that can be used in hydrological applications, such as PostGIS, GDAL/OGR, and many Python libraries, which are covered by other books from Locate Press.

On the IHE Delft OpenCourseWare site for GIS (http://www.gisopencourseware.org), you can find frequently updated learning resources for hydrological applications, including GDAL and Python tutorials. In addition, the videos on the YouTube channel (https://www.youtube.com/c/HansvanderKwast) show the latest developments.

There is a new QGIS release every four months and a new long-term release annually. Stay current on new features by browsing the Visual Changelogs at https://www.qgis.org/en/site/forusers/visualchangelogs.html

8.1 Things to Do

Learn more skills via the QGIS Training Manual: https://www.qgis.org/en/site/forusers/trainingmaterial/index.html.

Refer to the QGIS User Guide when technical questions arise: https://docs.qgis.org/3.22/en/docs/user_manual/.

Look on GIS.StackExchange.com for solutions to issues you encounter: https://gis.stackexchange.com/questions/tagged/qgis.

Subscribe to a QGIS Mailing List: https://www.qgis.org/en/site/getinvolved/mailinglists.html.

Read *How to ask a QGIS question* before posing your first question to a mailing list (listserv): https://www.qgis.org/en/site/getinvolved/faq/index.html#how-to-ask-a-qgis-question.

Read about how others have used QGIS in QGIS Case Studies: https://www.qgis.org/en/site/about/case_studies/index.html.

Explore maps others have made using QGIS in the Flickr Map Showcase: https://www.flickr.com/groups/qgis/pool/.

Read blog entries about new features and how-tos at http://plugins.qgis.org/planet/:

- Anita Graser: https://anitagraser.com/
- North Road: https://north-road.com/blog/
- Klas Karlsson: https://www.youtube.com/channel/UCxs7cfMwzgGZhtUuwhny4-Q

- Spatial Thoughts: `https://spatialthoughts.com/blog/`

Explore other QGIS Applications such as QGIS Server and QGIS Web Client: `https://www.qgis.org/en/site/about/features.html`

If you require commercial support for QGIS, you can go to this page to see companies providing a variety of support options: `https://www.qgis.org/en/site/forusers/commercial_support.html`.

We also hope that this book has inspired you to get involved as a QGIS community member. There are different ways to contribute to QGIS. You can help by reporting bugs, translating QGIS, developing plugins and new features, and sponsoring/donating to QGIS. Read more here: `https://www.qgis.org/en/site/getinvolved/index.html`.

Join your local user group or start one if one doesn't exist. Visit this page to see where there are established user groups: `https://www.qgis.org/en/site/forusers/usergroups.html`.

Visit the Open Source GeoSpatial Foundation (OSGeo) website, to stay abreast of other FOSS4G project news and conference announcements: `http://www.osgeo.org/`.

Index

179

Books from Locate Press

Be sure to visit locatepress.com for information on new and upcoming titles.

Discover QGIS 3.x

EXPLORE THE LATEST LONG TERM RELEASE (LTR) OF QGIS!

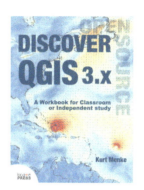

Discover QGIS 3.x is a comprehensive up-to-date workbook built for both the classroom and professionals looking to build their skills.

Designed to take advantage of the latest QGIS features, this book will guide you in improving your maps and analysis.

Discover QGIS 3.x is an update of the original title, using QGIS 3.6, covering Spatial analysis, Data management, and Cartography. The book includes new exercises and a new section—Advanced Data Visualization.

The book is a complete resource and includes: lab exercises, challenge exercises, all data, discussion questions, and solutions.

Leaflet Cookbook

COOK UP DYNAMIC WEB MAPS USING THE RECIPES IN THE LEAFLET COOKBOOK.

Leaflet Cookbook will guide you in getting started with Leaflet, the leading open-source JavaScript library for creating interactive maps. You'll move swiftly along from the basics to creating interesting and dynamic web maps.

Even if you aren't an HTML/CSS wizard, this book will get you up to speed in creating dynamic and sophisticated web maps. With sample code and complete examples, you'll find it easy to create your own maps in no time.

A download package containing all the code and data used in the book is available so you can follow along as well as use the code as a starting point for your own web maps.

QGIS Map Design - 2nd Edition

LEARN HOW TO USE QGIS 3 TO TAKE YOUR CARTOGRAPHIC PRODUCTS TO THE HIGHEST LEVEL.

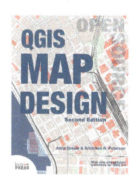

QGIS 3.4 opens up exciting new possibilities for creating beautiful and compelling maps!

Building on the first edition, the authors take you step-by-step through the process of using the latest map design tools and techniques in QGIS 3. With numerous new map designs and completely overhauled workflows, this second edition brings you up to speed with current cartographic technology and trends.

See how QGIS continues to surpass the cartographic capabilities of other geoware available today with its data-driven overrides, flexible expression functions, multitudinous color tools, blend modes, and atlasing capabilities. A prior familiarity with basic QGIS capabilities is assumed. All example data and project files are included.

Get ready to launch into the next generation of map design!

The PyQGIS Programmer's Guide
USE PYTHON TO EXTEND AND ENHANCE QGIS.

With PyQGIS you can write scripts and plugins to implement new features and perform automated tasks.

This book is updated to work with the next generation of QGIS—version 3.x. After a brief introduction to Python 3, you'll learn how to understand the QGIS Application Programmer Interface (API), write scripts, and build a plugin.

The book is designed to allow you to work through the examples as you go along. At the end of each chapter you will find a set of exercises you can do to enhance your learning experience.

The PyQGIS Programmer's Guide is compatible with the version 3.0 API released with QGIS 3.x and will work for the entire 3.x series of releases.

pgRouting: A Practical Guide
WHAT IS PGROUTING?

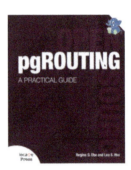

It's a PostgreSQL extension for developing network routing applications and doing graph analysis.

Interested in pgRouting? If so, chances are you already use PostGIS, the spatial extender for the PostgreSQL database management system.

So when you've got PostGIS, why do you need pgRouting? PostGIS is a great tool for molding geometries and doing proximity analysis, however it falls short when your proximity analysis involves constrained paths such as driving along a road or biking along defined paths.

This book will both get you started with pgRouting and guide you into routing, data fixing and costs, as well as using with QGIS and web applications.

Geospatial Power Tools
EVERYONE LOVES POWER TOOLS!

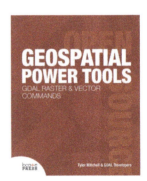

The GDAL and OGR apps are the power tools of the GIS world—best of all, they're free. The book is divided into three parts: Workflows and examples, GDAL raster utilities, and OGR vector utilities.

The utilities include tools for examining, converting, transforming, building, and analysing data. This book is a collection of the GDAL and OGR documentation, but also includes new content designed to help guide you in using the utilities to solve your current data problems.

Inside you'll find a quick reference for looking up the right syntax and example usage quickly. Once you get a taste of the power the GDAL/OGR suite provides, you'll wonder how you ever got along without them.

See these books and more at http://locatepress.com